DATE DUE	
7/15/98	4/28/98
# 210527?	8/23/99

SAMPLING, ALIASING, and DATA FIDELITY

for

Electronic Imaging Systems, Communications, and Data Acquisition

Gerald C. Holst

Copublished by

JCD Publishing
2932 Cove Trail
Winter Park, FL 32789

and

SPIE OPTICAL ENGINEERING PRESS

A publication of SPIE—The International Society for Optical Engineering
Bellingham, Washington USA

Library of Congress Cataloging-in-Publication Data

Holst, Gerald C.
  Sampling, aliasing, and data fidelity for electronic imaging systems, communications, and data acquisition/ Gerald C. Holst.
     p. cm.
  Includes bibliographical references and index.
  ISBN 0-9640000-3-2 (hardcover)
  1. Imaging systems 2. Signal processing. I. Title
  TK8315.H65 1997
  621.36'7--dc21   97-34944
                                                                                     CIP

Copublished by

JCD Publishing
2932 Cove Trail
Winter Park, FL 32789
Phone: 407/629-5370
Fax: 407/629-5370
ISBN: 0-9640000-3-2

SPIE- The International Society for Optical Engineering
P.O. Box 10
Bellingham, WA 98227-0010
Phone: 360/676-3290
Fax: 360/647-1445
WWW: http://www.spie.org
SPIE Volume PM-55
ISBN: 0-8194-2763-2

Notice:
  Reasonable efforts have been made to publish reliable data and information, but the Author and Publishers cannot assume responsibility for the validity of all materials or the consequences of their use.

Copyright © 1998 Gerald C. Holst

All rights reserved. No part of this book may be reproduced in any form by any means without written permission from the copyright owner.

This book is dedicated to my best friend

Cathy

PREFACE

The sampling theorem was developed for communications and is now applied to a variety of systems. When a band-limited signal is adequately sampled, it can be accurately reconstructed when an appropriate low-pass filter is present. Unfortunately, real signals are not usually band-limited and inefficient reconstruction filters are often used.

The mathematics associated with practical components and real signals often becomes cumbersome. Idealized components are used to keep the mathematics relatively simple and to allow easy graphical representations of signals. These idealizations are physically unrealizable and therefore lead to physically unrealizable results. However, the results are quite useful for drawing general conclusions. Frequently, the extension from real components to idealized components does not significantly affect the results.

While many electrical circuit designs approximately satisfy the sampling theorem, most electronic imaging systems are undersampled and this leads to aliasing. Aliasing is not always obvious when viewing complex scenery and is rarely reported during actual system usage although it is always present. It becomes apparent when viewing periodic objects such as test patterns, plowed fields, and Venetian blinds. Moiré patterns appear on these objects.

Sampling requirements are only one part of system design. Therefore, sampling cannot be studied in isolation. The detector array within the imaging system spatially samples the scene in two dimensions. The resultant time-varying signal is digitized by an analog-to-digital converter (ADC). Image processing manipulates the signal to enhance certain features. For an analog output, the signal passes through a digital-to-analog converter and a reconstruction filter. This output is often re-digitized by the data collection device (e.g., frame grabber or digitizing oscilloscope). If simply presented on a display, the display electronics will re-digitize the signal. Thus, when data are analyzed (by an observer or a computer), it has been spatially sampled (the detectors) and temporally sampled (the ADCs). This book covers both temporal and spatial sampling to ensure that a true end-to-end analysis is considered. It bridges the gap between sampling theory, signal appearance, and system performance.

A very important (and often overlooked) aspect of sampling theory is reconstruction. We cannot see digital data that reside in a computer memory. Most attempts to represent digital data include some form of a reconstruction filter. This filter is necessary to see digital TV, data sent to a digital computer monitor, halftoning in printing, and graphical representations. Often, the display

vi SAMPLING, ALIASING, and DATA FIDELITY

device is matched to the eye's spatial response to create a *perceived* continuous image. Although Chapter 7 discusses reconstruction filters in-depth, they are used in every chapter so that the digital data can be "seen". Machine vision systems do not have reconstruction filters. Here, the final "interpreter" is a computer algorithm that operates on the discrete data points.

Chapter 2 introduces the Fourier transform. It forms the basis of linear system theory (Chapter 3). Because a finite number of data points are available, the discrete Fourier transform is discussed. The finite number is equivalent to placing a window about an infinite set. By judiciously selecting a window size, the frequency components of a variety of test targets (e.g., three-bar and four-bar) can also be calculated.

The reconstructed analog signal, $v_{RECON}(t)$, may deviate from the original analog signal, $v(t)$, for several reasons. If $v(t)$ is not band-limited, aliasing occurs. This is easy to show in the frequency domain (Chapter 4). If aliasing occurs, $v_{RECON}(t)$ is a distorted version of $v(t)$. While an anti-alias filter can reduce aliasing, it may also affect signal fidelity. The sampling process creates new frequencies not present in the original signal. These are replications of the original signal that occur at multiples of the sampling frequency.

Sampling theory suggests that the signal be digitized by a flash ADC. The ADC samples and quantizes the signal. Detectors, on the other hand, also provide spatial integration (Chapter 5). The discrete location of the detectors creates the sampling lattice. Resampling changes the sampling frequency. When used with an imaging system, it creates samples at spatial locations other than where the original pixels were taken (Chapter 6).

If the reconstruction passes replicated frequencies, $v_{RECON}(t)$ is again distorted (Chapter 7). When the reconstruction filter attenuates in-band amplitudes, edges are softened and imagery appears blurry. For imaging systems, the overall reconstruction process may be a combination of electronic filters, the display medium, and the eye. Interpolation algorithms (used for resampling) and reconstruction filters are often similar in form and function.

While sampling theory applies to sinusoids, many signals are aperiodic. Phasing effects and edge location ambiguity are described in Chapter 8. Here, the difference between $v_{RECON}(t)$ and $v(t)$ is discussed. With bar patterns, the bars may appear to have different widths, different amplitudes, and a beat frequency envelope. These distortions do not violate the sampling theorem. They happen when the signal is not band-limited (aliasing occurs) and when an inefficient reconstruction filter is used.

Chapter 9 describes the various subsystem MTFs (modulation transfer function) so that the overall system MTF can be calculated. The electronics cannot "know" what created the signal nor the spectral content associated with the signal. The MTFs are generic to all systems.

Resolution measures were historically linked to the MTF. Resolution metrics now exist that include both the MTF and Nyquist frequency (Chapter 10). Chapter 11 provides several image quality metrics that apply to sampled data systems. Many metrics are based on the frequency domain differences of $v_{RECON}(t)$ and $v(t)$. They include the amount of aliased energy and spurious response. The metric selected depends on the application.

This book is for everyone working with digital signals. It describes how to represent digital signals on back-of-the-envelope drawings. This book is for the system analyst who wants to understand the implications of sampling theory. The system designer will maximize MTFs for some applications. In other applications, edge detection and signal width are important. This book provides both approaches. For the software specialist, this book provides the link between digital and analog data. Software cannot be effective if the sampling device characteristics and reconstruction process are not known. The display engineer and human factor specialist will understand why digital data can be seen.

This book does not contain digital image processing techniques. Many excellent texts exist. See, for example, *Fundamentals of Electronic Image Processing*, A. R. Weeks, Jr., SPIE Press (1996) or *The Imaging Processing Handbook*, 2nd edition, J. C. Russ, CRC Press (1995). As the number of quantization levels increase, the accuracy in representing the signal increases. The number required is application specific and therefore not described in detail.

The author extends his deepest gratitude to all his coworkers and students who have contributed to the ideas in this book. They are too many to mention by name. The author especially thanks all those who read draft copies of the manuscript: Glenn Boreman, University of Central Florida; Ronald Driggers, Science Applications International; Carl Halford, University of Memphis; Herb Huey, Boeing; Terrence Lomheim, Aerospace Corporation; Harold Orlando, Northrop Grumman; Mark Sartor, Xybion; Richard Vollmerhausen, Night Vision and Electronic Sensor Directorate; and Arthur Weeks, Jr., University of Central Florida. Although these reviewers provided valuable comments, the accuracy is the sole responsibility of the author. Damaris Ortiz provided the graphic arts.

Gerald C. Holst

TABLE OF CONTENTS

PREFACE v
SYMBOL LIST xiii

1. INTRODUCTION 1
   1.1. Sample data systems 3
   1.2. Sampling and quantization 5
   1.3. Reconstruction 12
   1.4. Aliasing 14
   1.5. Data acquisition 15
   1.6. Audio communications 15
   1.7. Electronic imaging systems 16
      1.7.1. Image enhancement 16
      1.7.2. Image restoration 17
      1.7.3. Aliased signal 17
      1.7.4. Moiré patterns 18
      1.7.5. Machine vision 19
   1.8. Computer simulations 19
   1.9. System modeling 20
   1.10. Scenels, pixels, datels, disels, and resels 24
   1.11. References 28

2. FOURIER TRANSFORM 29
   2.1. Fourier series 31
   2.2. Fourier integral 39
   2.3. Transform properties 42
   2.4. Discrete fourier transform 44
   2.5. Two-dimensional transform representation 51
   2.6. Windows 59
   2.7. Typical test patterns 62
   2.8. References 65

3. LINEAR SYSTEM THEORY 65
   3.1. Linear system theory 66
      3.1.1. Time-varying signals 67
      3.1.2. Spatially varying signals 69
   3.2. The electronic imaging system 70
   3.3. MTF and PTF interpretation 72
   3.4. Response of idealized circuits 77
   3.5. Superposition applied to optical systems 78
   3.6. References 80

x SAMPLING, ALIASING, and DATA FIDELITY

4. SAMPLING	81
4.1. Sampled data systems	82
4.2. Sampling theory	84
4.3. Reconstruction filter	87
4.4. Aliasing	89
4.5. Anti-alias filter	91
4.6. Two-dimensional sampling	92
4.7. References	97
5. SAMPLING DEVICES	98
5.1. Analog multiplexer	99
5.2. Analog-to-digital converters	102
5.3. Detectors	103
5.3.1. Detector MTF	103
5.3.2. Detector array output	107
5.3.3. Nyquist frequency	110
5.3.4. Microscan	113
5.3.5. Infinite staring arrays	115
5.3.6. Infinite scanning arrays	119
5.3.7. Finite arrays	122
5.3.8. Optical prefiltering	123
5.3.9. Nonrectangular sampling	126
5.3.10. Frequency axis normalization	127
5.4. Frame grabbers	129
5.5. References	130
6. RESAMPLING	131
6.1. Examples	132
6.2. Decimation	136
6.3. Interpolation	139
6.4. Frequency shift (rational value)	142
6.5. Interpolation algorithms	144
6.6. Frame grabbers	154
6.7. References	157
7. RECONSTRUCTION	158
7.1. Time domain reconstruction	161
7.2. The observer	168
7.3. Reconstruction by the display medium	171
7.3.1. Raster scanned CRTs	173
7.3.2. Digitally addressed CRTs	179
7.3.3. Color CRTs	179

| 7.3.4. Flat panel displays 181
| 7.3.5. Laser printers 183
| 7.3.6. Halftones 185
| 7.4. Electronic zoom 186
| 7.5. System aliasing 193
| 7.6. References 195

8. RECONSTRUCTED SIGNAL APPEARANCE 196
   8.1. Phasing effects 197
   8.2. Edge ambiguity 202
   8.3. Target width ambiguity 205
       8.3.1. Small blur-to-detector ratio 205
       8.3.2. Large blur-to-detector ratio 208
   8.4. Beat frequencies 209
   8.5. Bar target appearance 215
   8.6. Independent pixels on target 221
   8.7. Character recognition 224
   8.8. Asymmetric sampling 225
   8.9. Dynamic sampling 226
   8.10. References 228

9. SYSTEM ANALYSIS 228
   9.1. Frequency domain 230
   9.2. Optics MTF 235
   9.3. Detectors 237
   9.4. Sample-scene MTF 238
   9.5. Electronic analog filters 239
       9.5.1. Low-pass filter 240
       9.5.2. Boost 243
   9.6. Digital filters 245
   9.7. Display 249
       9.7.1. CRT-based display 249
       9.7.2. Flat panel display 251
   9.8. The observer 251
   9.9. System MTF 253
   9.10. Simulation of subsystem MTFs 260
       9.10.1. Computer generated imagery 260
       9.10.2. High resolution imagery 260
       9.10.3. Aliasing 261
       9.10.4. Number of detectors 263
       9.10.5. Frequency scaling 264
       9.10.6. Reconstruction 264

xii SAMPLING, ALIASING, and DATA FIDELITY

- 9.11. Dynamic scene projectors 269
- 9.12. References 273

- 10. SYSTEM RESOLUTION 274
  - 10.1. Electronic imaging system resolution 276
    - 10.1.1. Optical resolution metrics 277
    - 10.1.2. Detector resolution 278
    - 10.1.3. Electrical resolution metric 281
  - 10.2. CRT resolution 281
    - 10.2.1. Vertical resolution 281
    - 10.2.2. Theoretical horizontal resolution 282
    - 10.2.3. TV limiting resolution 282
  - 10.3. MTF-based resolution 283
    - 10.3.1. Limiting resolution 283
    - 10.3.2. MTF/TM resolution 283
    - 10.3.3. Séquin's limiting resolution 285
  - 10.4. Shade's equivalent resolution 286
  - 10.5. System resolution examples 288
  - 10.6. References 293

- 11. IMAGE QUALITY METRICS 293
  - 11.1. Image quality model 296
  - 11.2. MTF 300
  - 11.3. Equivalent pass band 301
  - 11.4. MTFA 302
  - 11.5. Subjective quality factor 302
  - 11.6. Square-root integral 305
  - 11.7. MRT and MRC 305
  - 11.8. Aliased signal 308
    - 11.8.1. Legault criterion 308
    - 11.8.2. Spurious response 308
    - 11.8.3. Aliased signal as noise 310
  - 11.9. Sampling and reconstruction blur 311
  - 11.10. NIIRS 311
  - 11.11. General trends 319
  - 11.12. References 320

- INDEX 323

SYMBOL LIST

a_n	Fourier series coefficient
a_o	Average value
ADC	Analog-to-digital converter
b_n	Fourier series coefficient
B	Monitor brightness
BW	Bandwidth
c_n	Fourier series coefficient
c/d/c	Continuous-discrete-continuous system
CCD	Charge-coupled device
CRT	Cathode ray tube
d_o	Period in space
d_x	Period in x direction
d_y	Period in y direction
d_{AIRY}	Airy disk diameter
$d_{ARRAY-H}$	Overall array size in horizontal direction
$d_{ARRAY-V}$	Overall array size in vertical direction
d_{CCH}	Horizontal center-to-center spacing
d_{CCV}	Vertical center-to-center spacing
d_H	Horizontal detector dimension
d_{H-FP}	Flat panel element horizontal dimension
d_{SPOT}	Distance between two spots on a digitally addressable display
d_{RASTER}	Distance between two raster lines on a scanned CRT
d_V	Vertical detector dimension
d_{V-FP}	Flat panel element vertical dimension
dpi	Dots per inch
D	Viewing distance
D_o	Aperture diameter
DAC	Digital-to-analog converter
DAS	Detector-angular-subtense
DAS_H	Horizontal detector-angular-subtense
DFT	Discrete Fourier transform
DN	Digital number
EIFOV	Effective instantaneous field-of-view
f	Frequency (Hz)
f_c	Filter cutoff
f_e	Electrical frequency
f_{eS}	Electrical sampling frequency
f_{e3dB}	Half power electrical frequency
f_o	Fundamental frequency of test target
f_n	Discrete frequency
f_v	Electrical frequency (video domain)
f_{vC}	Electrical cutoff frequency (video domain)
f_{v3dB}	Half power video frequency
f_{BEAT}	Beat frequency
f_{BOOST}	Boost frequency
f_{EIFOV}	Frequency where MTF = 0.5
f_H	Highest frequency present
f_N	Nyquist frequency
f_{N1}	Nyquist frequency: $f_{N1} = f_{S1}/2$
f_{N2}	Nyquist frequency: $f_{N2} = f_{S2}/2$
f_{PEAK}	Peak frequency of HVS response
f_S	Sampling frequency
f_{S1}	Input sampling frequency
f_{S2}	Output sampling frequency
f_{MAX}	Maximum frequency
f(t)	A time varying function
f(x,y)	Two-dimensional function that varies with distance
fl	Optic focal length
F	f-number
F_{3-BAR}	Transform of 3-bar target
F_{4-BAR}	Transform of 4-bar target
F(f)	Fourier transform of f(t)
F(u,v)	Fourier transform of f(x,y)
FFT	Fast Fourier transform
FOV	Field-of-view
g(x)	A function
g(x,y)	Two-dimensional function that varies with distance
G(u,v)	Fourier transform of g(x,y)

xiv SAMPLING, ALIASING, and DATA FIDELITY

$h(t)$	Impulse response in time	N_e	Equivalent pass band
$h(x,y)$	Impulse response in space	N_o	Number of datels
$h\{\ \}$	An operator	N_{AVE}	Number of samples averaged
$h_{RECON}(t)$	Reconstruction filter impulse response	N_{CYCLE}	Number of cycles
		$N_{DAS\text{-}H}$	Number of horizontal DASs
$h_{RECON}(x,y)$	Reconstruction filter spatial response	N_{DET}	Number of scenels across detector
$H(f)$	Frequency response of a circuit	N_H	Number of horizontal detectors
$H_{RECON}(f)$	Reconstruction filter frequency response	N_{PITCH}	Number of scenels across detector pitch
$H_{MONITOR}$	Monitor height	N_{PIXEL}	Number of pixels
$H_{RECON}(u,v)$	Reconstruction filter MTF	N_{SAMPLE}	Number of samples
$H(u,v)$	Spatial frequency response	N_{SCENEL}	Number of scenels
HFOV	Horizontal field-of-view	N_{TARGET}	Number of independent samples on target
HVS	Human visual system		
		N_{TV}	Resolution in TVL/PH
$i(x,y)$	Image	N_V	Number of vertical detectors
$i_D(x,y)$	Displayed image		
$I(u,v)$	Transform of $i(x,y)$	$o(x,y)$	Object
$I_{AVE}(u)$	Average of transformed Image	$O(u,v)$	Fourier transform of $o(x,y)$
IFOV	Instantaneous field-of-view	OCR	Optical character reader
$\Im(f)$	Imaginary part of the Fourier transform	OTF	Optical transfer function
		PAS	Pixel-angular-subtense
j	Square root of -1	PH	Picture height
		PSF	Point spread function
k	An integer	PTF	Phase transfer function
K	Maximum value of an integer		
		Q	Boost filter quality factor
LSB	Least significant bit		
LSF	Line spread function	$r_i(t)$	Interpolation algorithm in time, i = order
LSI	Linear-shift-invariant		
		$r_i(x)$	Interpolation algorithm in space, i = order
m	An integer		
m_S	Frequency multiplier	R	Detector responsivity
M	Maximum value of an integer	$R(\lambda)$	System spectral responsivity
M_t	Modulation threshold	R_{EQ}	Equivalent resolution
MRC	Minimum resolvable contrast	$R_{EQ\text{-}SYS}$	System equivalent resolution
MRT	Minimum resolvable temperature	$R_i(f)$	Frequency response of $r_i(t)$
		$R_i(u)$	Frequency response of $r_i(x)$
MTF	Modulation transfer function	$\Re(f)$	Real part of the Fourier transform
MTF_{POST}	MTFs after detector		
MTF_{PRE}	MTFs up to and including the detector		
		s	Samples per pixel
MTFA	Modulation transfer function area	$s(t)$	Sampling function in time
		$s(x,y)$	Two-dimensional sampling function
n	An integer		
n_S	Frequency divisor	sinc(z)	$\sin(\pi z)/(\pi z)$
N	Maximum value of an integer	S	CRT spot diameter at 50% amplitude

SYMBOL LIST xv

$S(f)$	Sampling function in frequency	v_{EYE}	Vertical observer spatial frequency
$S(u,v)$	2-D sampling function in frequency	v_i	Vertical image spatial frequency
		v_{iD}	Vertical detector cutoff
$S_{NOISE}(f_e)$	Noise power spectral density		
SNR	Signal-to-noise ratio		
SNR_p	Perceived SNR	v_{iN}	Vertical image Nyquist spatial frequency
SQF	Subjective quality factor		
SQRI	Square root integral	v_{iS}	Vertical image sampling spatial frequency
SR	Radiometric error		
		v_o	Target fundamental frequency
t	Time	v_{ob}	Vertical object spatial frequency
t_d	Time delay	v_{oN}	Vertical Nyquist frequency
t_{H-LINE}	Time to read one line	v_{rS}	Vertical raster sampling frequency
t_{S1}	Sample period ($t_{S1} = 1/f_{S1}$)		
t_{S2}	Sample period ($t_{S2} = 1/f_{S2}$)	v_N	Vertical Nyquist frequency
$t_{VIDEO-LINE}$	Active TV line time	v_S	Vertical sampling frequency
t'	Dummy time variable	$v_{AVE}(t)$	Averaged time-varying voltage
T	Sample period	$v_{SAMPLE}(t)$	Sampled time-varying signal
T_{CLOCK}	Clock period	$v'_{SAMPLE}(t)$	Sampled time-varying signal
TVL	TV lines	$v_{RECON}(t)$	Analog output after reconstruction filter
u	Frequency (cycles/sample)	V	Voltage amplitude
u_d	Horizontal display frequency	$V(f)$	Frequency response of $v(t)$
u_{dS}	Horizontal display sampling frequency	VFOV	Vertical field-of-view
		V_{MAX}	Maximum voltage
u_{EYE}	Horizontal observer spatial frequency	V_{MIN}	Minimum voltage
		$V_{SAMPLE}(f)$	Fourier transform of $v_{SAMPLE}(t)$
u_i	Horizontal image spatial frequency	$V_{RECON}(f)$	Fourier transform of $v_{RECON}(t)$
		$V_{AVE}(f)$	Fourier transform of $v_{AVE}(t)$
u_{iC}	Optical system cutoff		
u_{iD}	Horizontal detector cutoff	$w(n)$	Discrete window function
u_{iN}	Horizontal image Nyquist spatial frequency	$w(t)$	Discrete function in time
		$W_{MONITOR}$	Monitor width
u_{iS}	Horizontal image sampling spatial frequency	x	Distance
		x'	Dummy space variable
u_n	Discrete frequency (cy/sample)	$x(n)$	Discrete function
u_o	Target fundamental frequency	$X(k)$	Fourier transform of $x(n)$
u_{ob}	Horizontal object spatial frequency		
		y	Distance
u_{oN}	Horizontal Nyquist frequency	y'	Dummy space variable
u_N	Horizontal Nyquist frequency		
u_{RES}	Frequency defining resolution	z	A variable
u_S	Horizontal sampling frequency	z'	Dummy variable
v	Frequency (cycles/sample)		
$v(t)$	Time-varying voltage	$\delta(f)$	Dirac delta
v_d	Vertical display frequency	δ_1	In-band filter tolerance
v_{dS}	Vertical display sampling frequency	δ_2	Out-of-band filter tolerance
		Δf	Frequency interval

xvi SAMPLING, ALIASING, and DATA FIDELITY

Δf_n	Discrete frequency interval
Δt	Time interval
$\Delta t'$	Dummy time interval
Δu	Frequency interval
Δv	Frequency interval
Δx	Bar width
Γ	Light level parameter for MTF_{HVS}
λ	Wavelength
θ	Phase
$\theta(f)$	Phase transfer function (one-dimensional)
$\theta(u,v)$	Phase transfer function (two-dimensional)
θ_n	Phase at frequency f_n
σ_{SPOT}	1/e diameter of CRT spot
τ	Pulse width
$\tau_{MINIMUM}$	Minimum pulse width
ω	Radian frequency

SAMPLING, ALIASING, and DATA FIDELITY

for

**Electronic Imaging Systems,
Communications, and Data Acquisition**

1

INTRODUCTION

With the incredible flexibility of digital data processing, data can be easily manipulated, stored, transmitted, and outputted. Perhaps the most compelling reason for adopting digital technology is the fact that the quality of digital signals remains intact during storage and reproduction unless they are deliberately altered. As a result, nearly all electronic devices use some digital data processing.

The analog signal is acquired, digitized, and then quantized. Data acquisition systems are required to measure accurately voltage, signal rise time, or signal pulse width. Digital systems are used to communicate (e.g., digital television, cellular phones, fax machines, modems, and compact disks). The increasing popular World Wide Web is based upon digital data communications. Digital images may be created by scanning a document, direct sampling (CCD camera), or be computer generated. These create electronic images and may be called electronic imaging systems.

Basic to the design of all these systems is how many digital samples are necessary to reproduce accurately the signal with minimal distortion. The answer depends upon the final interpreter of the digital data. This is the ear for audio communication, the human visual system (HVS) for visual communication, or a computer for data analysis. Electronic images interpreted by a computer algorithm are called machine vision systems.

The sampling theorem, as introduced by Shannon[1] was applied to information theory. He stated that if a time-varying function, v(t), contains no frequencies higher than f_{MAX} (Hz), it is completely determined by giving its ordinates at a series of points spaced $1/2f_{MAX}$ sec apart. The original function can be reconstructed by using an ideal low-pass filter. To just satisfy Shannon's theorem, the sampling frequency, f_S, must be $2f_{MAX}$. Shannon's work is an extension of others[2] and the sampling theorem is often called the Whittaker-Shannon theorem.

Three conditions must be met to satisfy the sampling theorem. The signal must be band-limited, the signal must be sampled at an adequate rate, and a low-pass reconstruction filter must be present. When any of these conditions are not present, the reconstructed analog data will not appear exactly as the originals.

2 SAMPLING, ALIASING, and DATA FIDELITY

Signals can be undersampled or oversampled. Undersampling is a term used to denote that the input frequency is greater than the Nyquist frequency (which is defined as one-half the sampling frequency). It does not imply that the sampling rate is inadequate for any specific application. Similarly, oversampling does not imply that there is excessive sampling. It simply means that there are more samples available than that required by the Nyquist criterion.

The sampling theorem is often used to provide a "yes/no" answer to the complicated question of whether a system can reproduce a signal with minimal distortion. It does not quantify the penalty for undersampling or the benefits of oversampling. The quoted "two samples per critical dimension" represents a mathematical, idealized situation. It is appropriate only when detecting sinusoids.

Shannon's sampling theorem was developed for the digitization and reconstruction of sinusoids. However, most signals encountered with data acquisition systems and electronic imaging systems are not periodic. Any function can be decomposed into a series of sinusoidal frequencies. If these frequencies are below the Nyquist frequency, then that signal can be faithfully reconstructed. But aperiodic objects can consist of an infinite number of frequencies. Clearly some of these frequencies will be above Nyquist frequency. Any frequency above Nyquist frequency will appear as a lower frequency after reconstruction. This is aliasing and it distorts signals.

Sampled data systems cannot be discussed without considering reconstruction, It is the reconstruction process that converts digital data into an analog format. We cannot see digital data. Most attempts to represent digital data include some form of a reconstruction filter. A reconstruction filter is necessary to see digital TV, data sent to a digital computer monitor, and halftoning in printing. The reconstruction filter may be the intentional blurring created by the television or computer monitor or the human visual system response. Machine vision systems do not have reconstruction filters. The final "interpreter" of a machine vision system is a computer algorithm that operates on discrete data points. A machine vision system is not an "imaging" system in the traditional sense.

System design does not start from sampling theory but from the system requirements. Sampling cannot be studied in isolation. System design probably should start with the final interpreter of the data and use his requirements as the basis of design. Simply put, the design should exploit the interpreter's capability so that the final signal *appears* identical to the original.

For audio communications (e.g., telephone and digital compact disks), the human ear determines data fidelity. Because the ear is very sensitive to frequency distortions, it is important to maintain frequency fidelity. But the HVS is primarily influenced by spatial rather than frequency content. The brain "fills in" missing data and can correctly identify partially obscured objects and objects with missing parts (Figure 1-1). Here, contextual clues provide the missing data. Consider a sentence containing **for ezample** or **for cxample**. In both cases, you would probably interpret it as **for example**. Optical character recognition systems have to be quite sophisticated to correctly identify **for example**.

The symbols used in this book are summarized in the *Symbol List* (page xiii) which appears after the *Table of Contents*.

Figure 1-1. Though heavy advertising, we "learn" to decipher the logo. Even if partially obscured, it is still interpretable. Pepsi is a registered trademark of the Pepsi-Cola Company.

1.1. SAMPLE DATA SYSTEMS

Both sampling and quantization are necessary to produce digital data. Electronic circuits use analog-to-digital converters (ADCs) to perform these functions (Figure 1-2). Sampling provides data at discrete locations. It introduces ambiguity in edge location. Quantization (or digitization) affects the signal amplitude. Sampling is an inherent feature of all electronic imaging systems. The discrete locations of the detector elements spatially sample the scene (Figure 1-3). Because the detector output is usually in analog format, an ADC digitizes the signal.

4 SAMPLING, ALIASING, and DATA FIDELITY

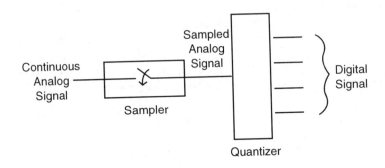

Figure 1-2. Analog-to-digital converter. The switch closes momentarily to sample the analog signal. A 4-bit ADC provides four outputs.

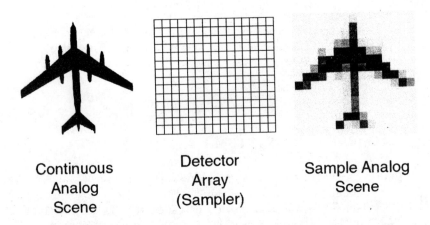

Figure 1-3. The detectors spatially sample the scene. They also spatially integrate the scene intensity to provide an average value. Each detector analog output is represented by a gray level here. After digitization, the scene becomes an array of digital numbers that reside in a computer memory.

1.2. SAMPLING and QUANTIZATION

Analog signals are continuous whereas sampled signals occur at discrete intervals. Quantization restricts the amplitude to specified levels. The digital signal can be translated into a series of "0"s and "1"s called binary digits or bits. The "0" and "1" represent two discrete states and may be called "off" and "on." Any other representation such as "a" and "b" could be used. There is nothing fundamental about using "0" and "1." Any system that can support discrete states can be used for digital data. For example, a low voltage could represent "0" and a high voltage could mean "1." Similarly, a bit value can be stored as a voltage on a chip, orientation of a magnetic spot on a disk or tape, or a pit and land on an optical disk. The minimum increment is the least significant bit (LSB). The total number of bits is usually an even number such as 8, 10, or 12. They provide 2^8, 2^{10}, and 2^{12} separate values, respectively. Each value is often called a gray level. While any number may be used, the 8-bit system is very popular. It provides 256 quantized levels.

Several codes, such as binary, octal, and hexadecimal, can be used to represent the digital value (Table 1-1). The analog signal is converted into base 2, base 8, and base 16 numbering systems, respectively. In this book, the decimal equivalent is used and labeled as a digital number (DN).

Figure 1-4 illustrates a time-varying analog signal that has been sampled every 0.1 time-units. The ideal sampler provides an output that is equal to the analog signal amplitude at the sampling time. An ADC also quantizes the analog voltages. In Figure 1-4b, the ADC range was selected so that one volt corresponds to a digital number of 7. The data will be quantized into eight levels (3 bits) whose digital numbers range from 0 to 7. A continuum of analog voltages is mapped into one digital number. For example, all voltages from 0.286 to 0.429 are mapped into 2 DN.

The output of an analog-to-digital converter is a series of digital numbers that are placed into a memory for further processing. While in the memory, the data are identified by memory locations or array indices. Table 1-2 illustrates 11 numbers that correspond to the digitized data shown in Figure 1-4b. A graphical representation helps in visualizing the data (Figure 1-5). Digital data have no further meaning until the user assigns units to the array indices. Array indices represent time for electronic circuits and space for imaging systems. The digital number may represent volts and intensity, respectively. A digital-to-analog converter changes the digital values into an analog voltage. Then a reconstruction filter effectively "connects" the data points (discussed in Section 1.3., *Reconstruction*).

Table 1-1
VARIOUS REPRESENTATIONS FOR A 4-BIT SYSTEM

DECIMAL	BINARY	OCTAL	HEXADECIMAL
0	0000	00	0
1	0001	01	1
2	0010	02	2
3	0011	03	3
4	0100	04	4
5	0101	05	5
6	0110	06	6
7	0111	07	7
8	1000	10	8
9	1001	11	9
10	1010	12	A
11	1011	13	B
12	1100	14	C
13	1101	15	D
14	1110	16	E
15	1111	17	F

INTRODUCTION 7

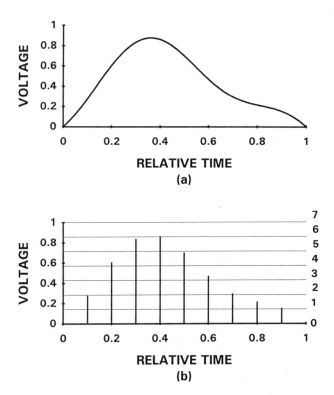

Figure 1-4. (a) Analog signal and (b) sampled time-varying signals. The sampled signal has the same amplitude as the analog signal at each sampling location. The scale at the right in (b) represents the digital values associated with a 3-bit system.

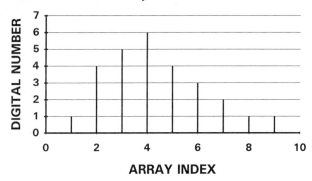

Figure 1-5a. Two methods of representing digital data. Continued on next page.

8 SAMPLING, ALIASING, and DATA FIDELITY

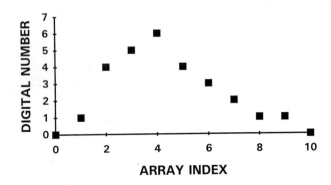

Figure 1-5b. Continued.

Table 1-2
3-BIT DATA ARRAY

ARRAY INDEX	DIGITAL NUMBER
0	0
1	1
2	4
3	5
4	6
5	4
6	3
7	2
8	1
9	1
10	0

INTRODUCTION 9

In Figure 1-6 the output data rate is identical to the digitizing data rate. Furthermore the output gain was selected such that 7 DN corresponds to one volt. Now the digital data can be compared directly to the original analog signal. Two important features of Figure 1-6 must be noted. The digitization into only eight levels forced the signal amplitude to vary considerably from the original signal. That is, digitization has introduced uncertainty into the actual signal level. The discrete locations of the sampled data do not permit an accurate representation of the time-varying signal.

Increasing the sampling rate (more samples per unit time) and the number of digital levels allows a better representation of the input signal (Figure 1-7). The number of bits required depends upon the application. More bits reproduce finer and finer amplitude detail. This may be important with medical and scientific imagery. But fax machines only need to detect the present of ink and therefore can be 1-bit systems (ink/no ink).

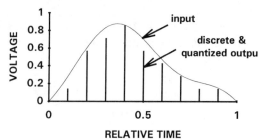

Figure 1-6. An insufficient number of samples and insufficient quantization introduce distortion. The digital data were clocked out every 0.1 time units and exist only at these discrete times. A reconstruction filter is necessary to make the data appear continuous.

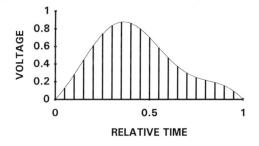

Figure 1-7. Increasing the sampling rate and number of digital levels allows good signal reproduction. The voltage has been quantized to 8 bits (256 levels). The difference between the signal and the digital level cannot be seen in this small graph. It always exists.

10 SAMPLING, ALIASING, and DATA FIDELITY

A difference exists between the actual signal and digitized signal. When expressed as a digital number, the average variance of the error is

$$VARIANCE = \frac{1}{12} \tag{1-1}$$

and the rms value is

$$RMS = \frac{1}{\sqrt{12}}. \tag{1-2}$$

When expressed as an analog signal,

$$v_{RMS} = \frac{LSB}{\sqrt{12}}, \tag{1-3}$$

where LSB is the voltage associated with one bit. The rms error is often called the quantization noise, and then the maximum signal-to-noise ratio is

$$SNR_{MAX} = \frac{v_{MAX}}{v_{RMS}} = \frac{2^N}{\frac{1}{\sqrt{12}}} = \sqrt{12}\, 2^N \tag{1-4}$$

or

$$SNR_{MAX\text{-}dB} = 20 \log(SNR) = 10.8 + 6.02N \quad dB. \tag{1-5}$$

The ADC's full scale is rarely used. That is, the actual signal is usually less than v_{MAX}. Here the SNR is

$$SNR = \frac{v}{v_{RMS}} = \sqrt{12}\, v_{DN}, \tag{1-6}$$

where v_{DN} is the digital number associated with v. As v decreases relative to v_{MAX}, the quantization errors illustrated in Figure 1-6 become more apparent. Simply stated, analog signals should be amplified so that v approaches v_{MAX} to obtain the full capability of the ADC.

INTRODUCTION 11

Sampling creates ambiguity in signal edge location. Apparent edge location depends upon the signal location with respect to the sampler. With data acquisition systems, it is the relative time between the signal and the ADC clock signal. For electronic imaging systems, it is the relative location of the object with respect to the detector locations.

Figure 1-8 illustrates the sampled data points as a function of phase (relative location of the signal with respect to the sampler). Table 1-3 provides the values. As the phase changes, the digital data changes. Increasing the sampling rate will minimize these variations and the associated artifacts.

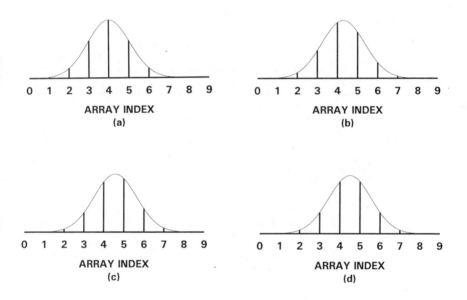

Figure 1-8. Digitized Gaussian pulse that appears different because of phasing effects. The vertical lines represent digital data.

12 SAMPLING, ALIASING, and DATA FIDELITY

Table 1-3
DIGITAL NUMBER (8-BIT SYSTEM)

Array index	Figure 1-8a	Figure 1-8b	Figure 1-8c	Figure 1-8d
0	0	0	0	0
1	4	2	0	1
2	43	31	13	15
3	163	139	85	93
4	255	251	220	228
5	163	187	235	228
6	43	57	102	93
7	4	7	18	15
8	0	0	1	1
9	0	0	0	0

1.3. RECONSTRUCTION

We cannot "see" digital data. The data become visible when "reconstructed." The reconstructed signal will be identical to the original signal when the signal is band-limited, the digitizer samples the signal at an adequate rate, and a low-pass reconstruction filter is present. Numerous algorithms exist that reconstruct the continuous analog signal.

Figure 1-9 illustrates a simple, back-of-the-envelope approach: the signal level is held constant from one digital data point to the next. This algorithm is the zero-order filter or sample-and-hold filter. It makes the data look "blocky." Figure 1-10 illustrates the same data but with a first-order or linear interpolation reconstruction filter. This filter "connects the dots." As the filter order increases, the analog image appears more like the original.

The ideal reconstruction filter will have unity amplitude up to the Nyquist frequency then sharply drop to zero. This removes the high frequency replicas created by the sampling process. The Gaussian curve is not band-limited and therefore cannot, in theory, be sampled adequately. The two reconstruction filters used in Figures 1-9 and 1-10 are not ideal low-pass filters. Some frequencies created by the sampling process remain in the reconstructed signal.

INTRODUCTION 13

These are spurious frequencies not in the original image. These frequencies are sometimes labeled as aliasing, but they are not. Aliasing is due to overlap. These frequencies are removed with better reconstruction filters (i.e., sharper cutoff filters).

Three different reconstruction filters can be used in an imaging system. The first is an electronic filter that band-limits the signal to the Nyquist frequency. The second reconstruction filter is the display medium. It purposely represents each digital datum as a finite spot. By overlapping the spots, the data appear continuous.

The final reconstruction filter is the HVS. By sitting at a normal distance from the display, the HVS spatial frequency response removes the higher order frequencies and the image appears continuous. The HVS's limited response makes printed halftone images appear continuous. The printing process can only provide ink dots. It cannot change the opacity. By placing ink dots close together, the HVS averages over many dots. As the dot density increases, the average value increases to give the appearance of increasing opacity.

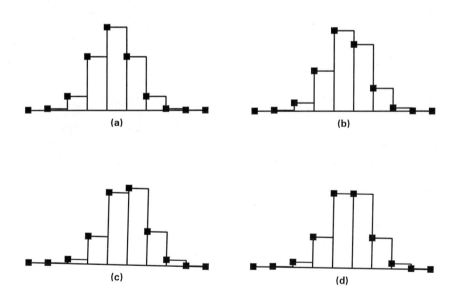

Figure 1-9. The zero-order sample-and-hold provides the typical back-of-the-envelope representation of digital signals. Reconstructed data from Figures 1-8a through 1-8d. The dots represent digital data.

14 SAMPLING, ALIASING, and DATA FIDELITY

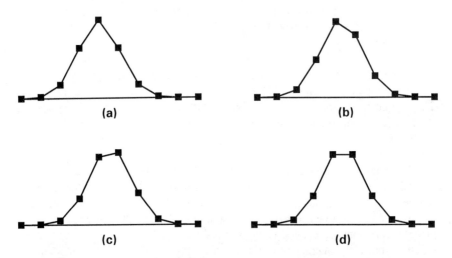

Figure 1-10. The first-order filter "connects the dots." Reconstructed data from Figures 1-8a through 1-8d. Many other reconstruction filters provide smoother analog signals.

1.4. ALIASING

The highest frequency that can be faithfully reconstructed is one-half the sampling rate. Any input signal above the Nyquist frequency, f_N, will be aliased down to a lower frequency. That is, an undersampled signal will appear as a lower frequency after reconstruction (Figure 1-11). After aliasing, the original signal can never be recovered. It would appear that aliasing is an extremely serious problem. However, the extent of the problem depends upon the final interpreter of the data.

Aliasing can be reduced to a negligible value (ideally eliminated) by using (1) a higher sampling frequency or (2) a low-pass anti-alias filter before the ADC to band-limit the signal. For data acquisition systems, technology limits the ADC sampling rate. With electronic imaging systems, the detector center-to-center spacing determines the sampling frequency. The ideal anti-alias filter will have unity response up to the Nyquist frequency and zero response after that. Unfortunately, real filters may affect the in-band (below Nyquist frequency) signal amplitude. Both the ideal anti-alias filter and ideal reconstruction filter have the same band pass characteristics. However, their purposes are different.

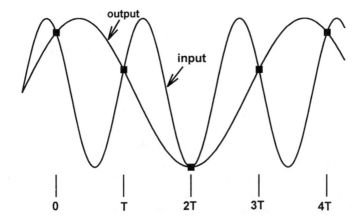

Figure 1-11. An undersampled sinusoid will appear as a lower frequency after reconstruction. The sampling frequency is $f_S = 1/T$. The input frequency is $1/1.43T$ and the output appears as $1/3.33T$. Nyquist frequency is $1/2T$. When T is measured in mrad, mm, or time, the sampling frequency is expressed in cycle/mrad, cycle/mm, or Hz, respectively.

1.5. DATA ACQUISITION

Data acquisition systems digitize analog signals. They are specified by the sampling rate and number of bits. Because data acquisition systems are not normally used to measure sinusoids, the issues associated with aliasing are not generally discussed in that literature. Data acquisition systems include frame grabbers and digitizing oscilloscopes.

Sampling creates ambiguity in edge location of one sample width. A pulse must be at least one sample wide to be recorded. Data fidelity centers on the ability to reproduce a pulse width or rise time.

1.6. AUDIO COMMUNICATIONS

The ear is sensitive to frequency distortions and efforts are made to prevent aliasing. Because the ear's frequency response for most people is limited to 20 KHz, the analog signal created by a microphone can be band-limited to 20 KHz with no loss in fidelity and then digitized at 40 KHz. To hear the signal (you cannot hear digital data), the signal must be reconstructed.

To conserve bandwidth, telephones signals are band-limited to about 3.5 KHz. This reduces the sound quality (cannot hear full symphony sounds) but is adequate for speech which generally contains lower frequencies. This represents a tradeoff between sampling rate and data fidelity.

1.7. ELECTRONIC IMAGING SYSTEMS

Sampling is an inherent feature of all electronic imaging systems. The discrete locations of the detector elements spatially sample the scene in both directions. Nearly all images are undersampled by the detector lattice. Because the detector spatially samples the scene, the scene must be band-limited optically to avoid aliasing. Spatial sampling creates ambiguity in target edge location and produces moiré patterns when viewing periodic targets. While this is a concern to scientific and military applications, it typically is of little consequence to the average consumer.

While aliasing is never considered desirable, it is accepted in scientific and military monochrome applications where high MTF (modulation transfer function) values are desirable. High MTF is related to image sharpness. Aliasing becomes bothersome when scene geometric properties must be maintained as with mapping. Aliasing affects the performance of most image processing algorithms. For specific applications such as medical imaging, it may be considered unacceptable. Here, an aliased signal may be misinterpreted as a medical abnormality that requires medication, hospitalization, or surgery.

Two significantly different image processing algorithms may exist within an electronic imaging system. They are image enhancement and image restoration. The success of both depends upon the reconstruction filter and the criterion used to evaluate image quality. Every image processing algorithm is optimized for a limited set of scenes. As the set size increases the algorithm is called robust. A scene can always be found where the algorithm provides poor results.

1.7.1. IMAGE ENHANCEMENT

Image enhancement converts the image into a form more amenable to human or machine analysis. No conscious effort is made to improve image quality or fidelity. Accentuated edges may provide a "better" image. Enhancement is simply called image processing.

Image enhancement is the manipulation of the data matrix to achieve a particular objective. This may include noise removal, geometric correction, radiometric correction, remapping the image onto another scale, etc. Radiometric correction usually affects only the pixel values, whereas geometric correction requires interpolation. Numerous image processing algorithms exist and many texts are available.[3]

Some images belong to a small precious data set (e.g., remote sensing imagery from interplanetary probes). They must be processed repeatedly to extract every piece of information. Some are part of a data stream that are examined once (e.g., real-time video) and others have become popular and are used routinely as standards. These include the three-bar or four-bar test patterns, Lena,[4] and the mandrill.

1.7.2. IMAGE RESTORATION

Image restoration is a process significantly different from reconstruction. It attempts to undo degradations that occur during the creation, processing, and transmission of an image. Atmospherics, sensor nonlinearities, noise, image blurring, and image motion can degrade imagery. If the source of degradation is known, then the image can often be restored. Simply, image restoration provides an estimate of the original scene from its degraded image. There exists literally hundreds of image restoration algorithms.[5] Their effectiveness depends, in part, upon the reconstruction filter characteristics. As a result, image restoration and image reconstruction are treated together in most literature. Imagery can be restored[6] when linear motion, defocus, and atmospheric blurring are present. Sampling effects can be minimized with Wiener-type reconstruction filters.[7] Restoration becomes difficult when the signal-to-noise ratio is low.

1.7.3. ALIASED SIGNAL

An electronic imaging system cannot perfectly reproduce the scene. The array spatially samples the image and the detector injects noise. The Nyquist frequency defines how much aliased signal and noise are present. Aliased signal effects on image quality have not been fully quantified.[8] Park and Hazra[9] treat the aliased signal as noise. Although they used only one image, they concluded ... *in this particular case, aliasing, not random noise ultimately limits the amount of restoration.* The amount of aliasing present is scene dependent. Optimized systems can be designed when the scene content is known.

18 SAMPLING, ALIASING, and DATA FIDELITY

1.7.4. MOIRÉ PATTERNS

Moiré patterns are produced when viewing periodic structures (Figure 1-12). Periodic structures are rare in nature and aliasing is seldom reported when viewing natural scenery although aliasing is always present. It may become apparent when viewing periodic targets such as picket fences, plowed fields, railroad tracks, and Venetian blinds. Aliasing becomes obvious when (a) the image size approaches the detector size and (b) the detectors are in a periodic lattice (the normal situation). Spatial aliasing is rarely seen in photographs or motion pictures because the grains are randomly dispersed in the photographic emulsion.

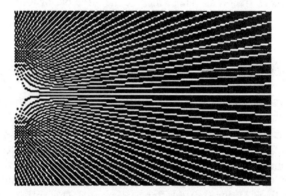

Figure 1-12. A raster scan system creates moiré patterns when viewing wedges or star bursts.

We have become accustomed to the aliasing in commercial televisions. Periodic horizontal lines are distorted due to the raster. Cloth patterns such as herringbones and stripes also produce moiré patterns. Cross-color effects occur in color imagery. Many video tape recordings are undersampled to keep the price modest and yet the imagery is considered acceptable normal viewing distances.

In a series of experiments performed by Barbe and Campana,[10] they conclude *Indeed, the prefiltering required to reduce the response beyond Nyquist appears to do more harm than good ..... the moiré pattern extends up to, but not through, the objects of interest; i.e., moiré in the display is limited to the extent of the scene causing the moiré.* It cannot exist elsewhere in the displayed image. The viewer is primarily influenced by the spatial content of the reproduction rather than the frequency content.

1.7.5. MACHINE VISION

A machine vision system consists of a camera and a computer. The software rapidly analyzes digitized images with respect to object location, size, flaws, and other preprogrammed data. Unlike other types of image analysis, a machine vision system also includes a mechanism that immediately reacts to images that do not conform to the parameters stored in the computer. For example, defective parts are removed from a production line conveyor belt. While a machine vision system does not require a monitor, a monitor is often used during system setup and for diagnostic evaluation.

Because machine vision systems are not normally used to measure sinusoids, the issues associated with aliasing are not generally discussed in machine vision literature. Rather, machine vision research focuses on the accuracy of detecting object features. Accuracy is related to the number of pixels on the feature and this becomes the metric used most often.

1.8. COMPUTER SIMULATIONS

Computer generated imagery consists of a two-dimensional array of numbers residing in a memory. The data may be displayed on a monitor where there is one-to-one mapping from the data array to the monitor "pixel." The monitor and HVS act as the reconstruction filters.

With hardware-in-the-loop-testing,[11] computer generated scenes are projected onto an imaging system or machine vision system. The number of scene elements should be much greater than the number of detectors to avoid additional sampling and phasing effects.

Similarly, with system performance simulations, the number of scene elements should be much greater than the number of detectors. This allows the simulation to account for the spatial integration provided by the detectors adequately. As a rule of thumb, there should be at least 16 scene elements per detector element.

1.9. SYSTEM MODELING

Following the approach proposed by Park[8], five different system models exist:

CONTINUOUS/CONTINUOUS (c/c) MODEL

This model simulates the performance of optical instruments such as telescopes and film-based cameras (Figure 1-13). It includes data acquisition devices such as voltmeters whose output is displayed on a meter. Many image quality metrics are based upon this approach. Radio was originally a c/c system. However, the components in both the transmitter and receiver have been replaced with digital electronics. The transmission is still analog. Similarly, television was c/c in the horizontal direction. The raster pattern in the vertical direction is a sampling lattice.

DISCRETE/DISCRETE (d/d) MODEL

Computer simulations are based upon this model. Both the input and output are computer generated. The input image is assumed to be a near-perfect replica of some scene without regard to optical blurring or the spatial integration afforded by the detector. Image processing algorithms, as presented in many texts use the d/d model (Figure 1-14). To display the results, the discrete data must be transformed into a continuous image Because we cannot see data residing in a computer memory.

CONTINUOUS/DISCRETE MODEL

Figure 1-15 illustrates the c/d system. The continuous input is digitized and then processed. Machine vision systems are c/d systems.

INTRODUCTION 21

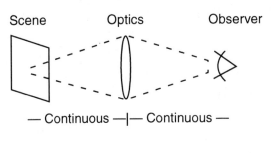

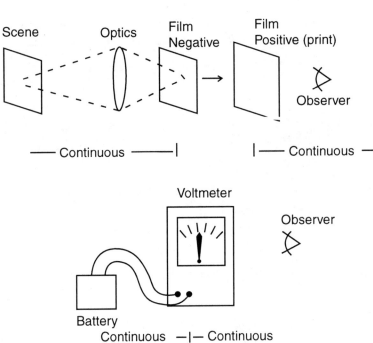

Figure 1-13. Optical systems and film-based imagery are continuous/continuous systems. Measuring equipment with meters are typically analog devices.

22 SAMPLING, ALIASING, and DATA FIDELITY

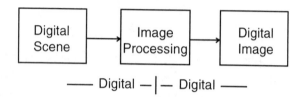

Figure 1-14. The discrete/discrete model. The input and output are data arrays residing in a computer memory. Any attempt to draw the image would make it continuous.

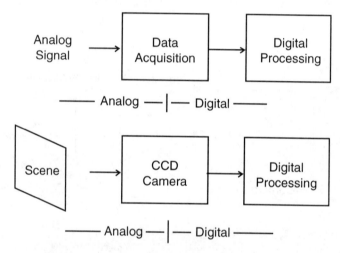

Figure 1-15. The continuous/discrete model.

CONTINUOUS/DISCRETE/CONTINUOUS (c/d/c) MODEL

Electronic imaging systems are described by this model (Figure 1-16). The model accounts for the optical blurring, detector spatial integration, sampling effects by the detector, all image processing algorithms, and then the conversion of the sampled image into either a viewable hard copy or soft copy. Each display medium has its own method of reconstructing the image. A sampled image will appear different on a computer monitor, laser printer, and film writer. For data acquisition systems, the signal amplitude is displayed on a read out in a "digital" format (e.g., numbers). The final interpreter of image quality is the observer.

INTRODUCTION 23

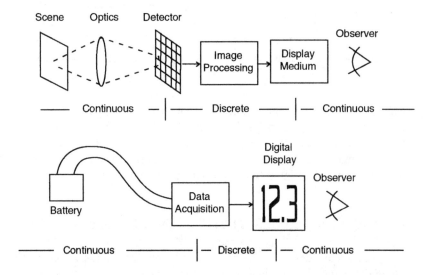

Figure 1-16. All electronic imaging systems are continuous/discrete/continuous. "Digital" displays are actually continuous devices whose output represents digital numbers (e.g., digital clock). They are controlled digitally.

CONTINUOUS/DISCRETE/CONTINUOUS/DISCRETE/CONTINUOUS (c/d/c/d/c) MODEL

Figure 1-17 illustrates a more complicated system with two samplers. Resampling may introduce additional aliasing and ambiguity in edge location. The output may be displayed on a variety of devices such as a CRT-based monitor, flat panel monitor, laser printer, or film writer. Although shown as an analog link, the data link could be digital. This figure also applies to conventional and high definition televisions. For remote sensing applications, digital imagery is resampled onto an earth-center coordinate system. This may include image rotation and image rectification.

24 SAMPLING, ALIASING, and DATA FIDELITY

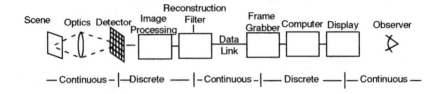

Figure 1-17. In many applications, the camera analog output is digitized for computerized data analysis. The display may be either hard copy (e.g., laser printer) or soft copy (CRT or flat panel display). The link may be simply a cable (wire) or a telecommunication channel. As shown the link is analog. Digital links may add more complexity.

1.10. SCENELS, PIXELS, DATELS, DISELS, and RESELS

Sampling theory describes the requirements that lead to the reconstruction of a digitized signal. The original analog signal is digitized, processed, and then returned to the analog domain. For electrical signals, each digitized value is simply called a sample and the digitization rate is the sampling rate.

Electronic imaging systems are more complex and several sampling lattices are present. Computed generated scenes (if used) are specified by the data array size. The detector output represents a sampling of the scene. The detector output is digitized and placed in a memory. After image processing, the data are output to a display medium. Although the display medium provides an analog signal, it is typically digitally controlled.

Each device has its own minimum sample size or primary element. Calling a sample a pixel or pel (picture element) does not seem bad. Unfortunately, there is no *a priori* relationship between the various device pixels. The various digital samples in the processing path are called scenels, pixels, datels, and disels (sometimes called dixels). For analog systems, the minimum size is the resel (Table 1-4). Digital "-els" are clearly defined and each array is mapped onto the next (Figure 1-18). However, there is no standard definition for a resel. For optical systems it may be the Airy disk or Rayleigh criterion. For electronic circuits, it is related to the bandwidth but the definition has not been standardized. When a conversion takes place between the analog and digital domains, the resel may be different from the digital sample size. The analog signal rather than the digital sample may limit the system resolution. In oversampled systems, the resel consists of many samples.

INTRODUCTION 25

Table 1-4
THE "-ELS"

ELEMENT	DESCRIPTION
Scenel (Scene element)	A sample created by a scene simulator. Because the data resides in a computer memory, the array size is equal to the number of scenels.
Pixel or pel (picture element)	A sample created by a detector.
Datel (data element)	Each datum is a datel. Datels reside in a computer memory.
Disel (display element)	The smallest element (sample) that a display medium can access.
Resel (resolution element)	The smallest signal supported by an analog system.

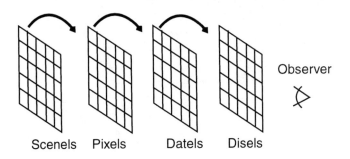

Figure 1-18. Each array is mapped onto the next. The number of elements in each array may be different. Not every array exists in every system.

26 SAMPLING, ALIASING, and DATA FIDELITY

For staring arrays, the total number of pixels is equal to the number of detectors. For scanning arrays, the number is determined by the ADC sampling rate and the number of detectors in the cross scan direction. The detector's spatial response is determined by the detector's size (e.g., photosensitive area). If the extent is d_H in the horizontal direction and the optics focal length is fl, then the horizontal detector-angular-subtense (DAS_H) is

$$DAS_H = \frac{d_H}{fl}. \tag{1-7}$$

Similarly, in the vertical direction,

$$DAS_V = \frac{d_V}{fl}. \tag{1-8}$$

Staring arrays are often specified by the detector center-to-center spacing (pitch). The horizontal pixel-angular-subtense (PAS_H) is

$$PAS_H = \frac{d_{CCH}}{fl} \tag{1-9}$$

where d_{CCH} is the horizontal pitch. Similarly, in the vertical direction

$$PAS_V = \frac{d_{CCV}}{fl}. \tag{1-10}$$

The fill-factor is the ratio of areas:

$$Fill\text{-}factor = \frac{d_H d_V}{d_{CCH} d_{CCV}}. \tag{1-11}$$

With a staring array that has a 100% fill-factor, the PAS is equal to the DAS (e.g., $d_H = d_{CCH}$ and $d_V = d_{CCV}$). They are also equal for a scanning array whose ADC provides one sample per DAS. Otherwise, they are different.

INTRODUCTION 27

If the Airy disk (a resel) is much larger than the PAS, then the optical resel determines the system resolution. If the number of pixels across this resel is large, the sampling theorem can be satisfied. If the electronic imaging system output is in a digital format, then the number of datels (samples) equals the number of pixels. If the camera's analog output is digitized, then the number of datels is linked to the frame grabber's digitization rate. This number can be much greater than the number of pixels. This higher number does not create more resolution in terms of the "-els." However, the higher sampling frequency may improve the overall system MTF.

Image processing algorithms operate on datels. A datel may not represent a pixel or a resel. Because there must be two samples to define a frequency, the Nyquist frequency associated with scenels, pixels, datels, and disels may be different. The discrete Fourier transform (discussed in Section 2.4., *Discrete Fourier Transform*) operates on datels. If there are more datels than pixels, the Nyquist frequency associated with the discrete Fourier transform is higher than that for the detector array. Most image processing books illustrate datels and call them pixels. The image processing specialist must understand the differences between the sampling lattices and who is the final data interpreter.

After image processing, the datels are output to a display medium. For monitors, each datel is usually mapped, one-to-one, onto each disel. Monitors are often specified by the number of addressable pixels (defined as a disel in this text). Although the number of addressable pixels (disels) may be large for cathode ray tube based monitors, the monitor resel may be limited by the electron beam diameter. With most display media, the finite-sized spots of two adjacent disels overlap to provide an analog (continuous) image.

Finally, the system designer must be aware of which subsystem limits the overall system resolution. In some respects, the starting point for system design should begin with the final interpreter of the data. The minimum "-el" should be discernible by the interpreter to ensure maximum transfer of information. However, the observer may not find this image aesthetically pleasing.

1.11. REFERENCES

1. C. E. Shannon, "Communication in the Presence of Noise," *Proceedings of the IRE*, Vol. 37, pp. 10-21 (January 1949).
2. A. B. Jerri, "The Shannon Sampling Theorem - Its Various Extensions and Applications: A Review," *Proceedings of the IEEE*, Vol. 85(11), pp. 1565-1595 (1977).
3. See, for example, *Fundamentals of Electronic Image Processing*, A. R. Weeks, Jr., SPIE Press (1996) or *The Imaging Processing Handbook*, 2nd edition, J. C. Russ, CRC Press (1995).
4. J. Van Tassel, "The Legend of Lena," *Advanced Imaging*, Vol. 11(5), pp. 56-60, (1996).
5. M. I. Sezan and A. M. Tekalp, "Survey of Recent Developments in Digital Image Restoration," *Optical Engineering*, Vol. 29(5), pp. 393-404 (1990).
6. Y. Yitzhaky and N. S. Kopeika, "Identification of the Blur Extent from Motion Blurred Images," in *Infrared Imaging Systems: Design, Analysis, Modeling, and Testing VI*, G. C. Holst, ed., SPIE Proceedings Vol. 2470, pp. 2-11 (1995).
7. J. A. McCormick, R. Alter-Gartenberg, and F. O. Huck, "Image Gathering and Restoration: Information and Visual Quality," *Journal of the Optical Society of America A*, Vol. 6(7), pp. 987-1005 (1989).
8. S. K. Park, "Image Gathering, Interpolation and Restoration: A Fidelity Analysis," in *Visual Information Processing II*, F. O. Huck and R. D. Juday, eds., SPIE Proceedings Vol. 1705, pp. 134-144 (1992).
9. S. K. Park and R. Hazra, "Image Restoration Versus Aliased Noise Enhancement," in *Visual Information Processing III*, F. O. Huck and R. D. Juday, eds., SPIE Proceedings Vol. 2239, pp. 52-62 (1994).
10. D. F. Barbe and S. B. Campana, "Imaging Arrays Using the Charge-Coupled Concept," in *Image Pickup and Display, Volume 3*, B. Kazan, ed., pp. 245-253, Academic Press (1977).
11. See, for example, *Technologies for Synthetic Environments: Hardware-in-the-loop Testing*, R. L. Murrer, Jr., ed., SPIE Proceedings Vol. 2741 (1996).

2
FOURIER TRANSFORM

Electronic imaging systems consist of components that respond to spatial variations (optics and detector) and components that respond to changes in time-varying electronic signals. Optical elements do not generally change with time and therefore are characterized only by spatial coordinates. Electronic circuitry is not sensitive to spatial activity. The detector provides the interface between the spatial and time-varying electronic components. The detector's response depends on both temporal and spatial quantities. The conversion of two-dimensional optical information to a one-dimensional electrical response assumes a linear photodetection process. Implicit in the detector response is the conversion from input flux to output voltage.

Two different temporal coordinates exist: "t'" applies to moving objects and "t" describes the electronic temporal signal. Because the electrical circuitry is not influenced by spatial changes, response can be separated from the spatial changes:

$$h_{SYSTEM}(x,y,t',t) = h_{SPATIAL}(x,y,t') \, h_{ELECTRONICS}(t). \qquad (2\text{-}1)$$

For stationary scenes (assumed throughout this text):

$$h_{SPATIAL}(x,y,t') = h_{SPATIAL}(x,y). \qquad (2\text{-}2)$$

The subscripts, SPATIAL and ELECTRONICS, will be dropped for brevity. The meaning is clear from the equations.

The usual approach is to develop the mathematics for the one-dimensional waveform and then expand it to two dimensions. This is convenient because one-dimensional functions are easy to portray graphically. Two-dimensional functions require a three-dimensional graph (two dimensions plus amplitude) and are more difficult to illustrate fully. Time-varying waveforms (e.g., voltage) use the one-dimensional transform. Objects, lens systems, and detector responses are described by the two-dimensional transform.

Time filters are different from spatial filters in two ways. Time filters are single-sided in time and must satisfy the causality requirement that no change in the output may occur before the application of an input. Optical filters are double-sided in space. Electrical signals may be either positive or negative, whereas optical intensities are always positive. As a result, optical and circuit designers use different terminology.

The Fourier series and Fourier transform are mathematical tools that relate a function to its frequency components. The Fourier series assumes that the signals are infinitely repetitive. Real signals have finite length and the discrete Fourier transform (DFT) is used. Because DFT software is readily available, the actual transform computation will not be discussed. The fast Fourier transform (FFT) is commonly used and is a computationally efficient method of computing the DFT.

When describing either optical or electronic system responses, it is convenient to select signals that are symmetrical about the origin. Then $f(-x) = f(x)$ and these signals are said to be "even." Some mathematical texts assume that the pulse starts at the origin so that it is "odd" or $f(-x) = -f(x)$. Even and odd functions provide different, but equivalent, expressions for the Fourier series. The difference is simply the choice of the origin. While either approach is valid, selecting an even function usually simplifies the mathematics.

The Fourier transform is first introduced in one dimension. The variable "t" is independent and "f" is the frequency variable. For the two-dimensional situation, "x" and "y" are the independent variables and "u" and "v" are the transformed variables. The variables f, u, and v have units of cycles per sample interval. This choice follows the common nomenclature where time (one-dimensional) is used to denote electronic circuitry response and space (x,y) is two-dimensional. Many texts transform data into radians per sample interval ($\omega = 2\pi f$). Although identical in meaning, equations may *appear* to differ by the factor 2π.

Lower case letters, e.g., f(), represent the independent function and upper case letters, e.g., F(), represent the transformed function. When used for electrical circuit analysis, $|H(f)|$ becomes the electronic transfer function and $\theta(f)$ is the phase transfer function. In the two-dimensional case, $H(u,v)$ is the complex optical transfer function (OTF). $|H(u,v)|$ is the optical MTF and $\theta(u,v)$ is the optical PTF. The combinations, $|H(u,v)||H(f)|$ and $\theta(u,v)\theta(f)$, characterize electronic imaging system performance. Further interpretation is provided in Chapter 3, *Linear System Theory*.

The symbols used in this book are summarized in the *Symbol List* (page xiii) which appears after the *Table of Contents*.

2.1. FOURIER SERIES

Any periodic function may be expanded into a Fourier series. For simplicity, let the function be one-dimensional and periodic in time with period T.

$$f(t) = \frac{a_o}{T} + \frac{2}{T}\sum_{n=1}^{\infty}\left[a_n\cos(2\pi f_n t) + b_n\sin(2\pi f_n t)\right], \qquad (2\text{-}3)$$

where

$$f_n = \frac{n}{T} \quad \text{where } n = 1, 2, \cdots \qquad (2\text{-}4)$$

The waveform's average value, a_o, is

$$a_o = \int_{-\frac{T}{2}}^{\frac{T}{2}} f(t)\, dt. \qquad (2\text{-}5)$$

The coefficients, a_n and b_n, are calculated from

$$a_n = \int_{-\frac{T}{2}}^{\frac{T}{2}} f(t)\cos(2\pi f_n t)\, dt \quad n = 1, 2, \cdots, \qquad (2\text{-}6)$$

and

$$b_n = \int_{-\frac{T}{2}}^{\frac{T}{2}} f(t)\sin(2\pi f_n t)\, dt \quad n = 1, 2, \cdots. \qquad (2\text{-}7)$$

In the complex exponential form, the Fourier series is

$$f(t) = \frac{1}{T}\sum_{n=-\infty}^{\infty} c_n e^{j2\pi f_n t}, \qquad (2\text{-}8)$$

where

$$c_n = a_n - jb_n = \int_{-\frac{T}{2}}^{\frac{T}{2}} f(t) e^{-j2\pi f_n t} dt \qquad (2-9)$$

and $j = \sqrt{-1}$. Some texts use i for $\sqrt{-1}$. Electrical engineering texts use j to avoid confusion with i which represents electrical current. (The reader may quickly surmise that the author was trained as an electrical engineer). The coefficient, a_n is the real part of c_n, and b_n is the imaginary part. The magnitude is

$$|c_n| = \sqrt{a_n^2 + b_n^2} . \qquad (2-10)$$

The phase is

$$\theta_n = \tan^{-1} \frac{b_n}{a_n} . \qquad (2-11)$$

Because $c_o = a_o$, substitution provides

$$f(t) = \frac{1}{T} \sum_{n=-\infty}^{\infty} c_n e^{j2\pi f_n t} = \frac{1}{T}\left(a_o + 2\sum_{n=1}^{\infty} |c_n| \cos(2\pi f_n t + \theta)\right). \qquad (2-12)$$

Plotting $|c_n|$ as a function of f_n provides the amplitude spectrum (also called the frequency spectrum). Plotting θ_n versus f_n is the phase spectrum. The square of the amplitude, $|c_n|^2$, presents power.

As an illustration, consider an infinite square wave as shown in Figure 2-1. By centering one pulse on the origin, the series becomes even. This means that only the cosine series exists ($b_n = 0$). The fundamental frequency is $f_o = 1/T$.

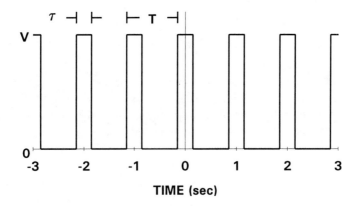

Figure 2-1. An infinite square wave whose period is T and pulse width is τ. The amplitude is V. This is an even function: f(-t) = f(t).

Expanding in f_n and solving for c_n provides

$$c_n = \frac{V}{\pi f_n} \sin(\pi \tau f_n) . \tag{2-13}$$

Rewriting provides

$$c_n = V\tau \frac{\sin(\pi \tau f_n)}{\pi \tau f_n} , \tag{2-14}$$

defining z as τf_n,

$$c_n = V\tau \frac{\sin(\pi z)}{\pi z} . \tag{2-15}$$

Because the function, $\sin(\pi z)/\pi z$, appears frequently,[1] it has been given a shorthand notation of sinc(z). Hereafter it will simply be called the sinc function. Using this shorthand notation,

$$c_n = V\tau \, sinc(\tau f_n) . \tag{2-16}$$

34 SAMPLING, ALIASING, and DATA FIDELITY

Because $f_n = n/T$,

$$c_n = V\tau \, sinc\left(\frac{n\tau}{T}\right). \tag{2-17}$$

As shown in Figure 2-2, the DC value exists because the signal shown in Figure 2-1 has an average value. The envelope of c_n is the sinc function. Figure 2-2 illustrates the normalized amplitude spectrum when $\tau \ll T$ and Figure 2-3 illustrates the spectrum for a square wave ($\tau = T/2$):

$$c_n = \frac{VT}{2} sinc\left(\frac{n}{2}\right). \tag{2-18}$$

Equivalently,

$$c_n = \left(\frac{VT}{2}\right) \frac{2}{\pi n} \sin\left(\frac{\pi n}{2}\right). \tag{2-19}$$

When n is even, $c_n = 0$. When n is odd, $\sin(\pi n/2)$ alternates between 1 and -1.

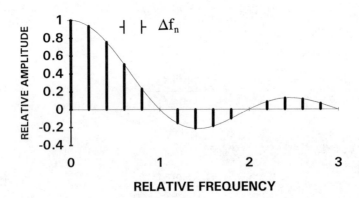

Figure 2-2. Normalized amplitude spectrum as a function of $n\tau/T$. The envelope is the sinc function. The sinc function is zero when $n\tau/T = 1$, 2, $\cdots$, . The frequency increments are τ/T or $\Delta f_n = \tau/T$. The amplitude has been normalized to $V\tau$. Only positive frequencies are shown ($n \geq 0$). The coefficient c_n is symmetrical about $n = 0$. The average value is also included ($f = 0$).

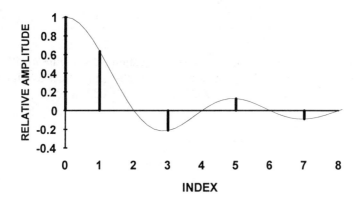

Figure 2-3. Normalized square wave amplitude spectrum as a function of n. Note that only the odd harmonics exist (n = 1, 3, ···). For even values (n = 2, 4, ···), sinc is zero. The amplitude has been normalized to VT/2. Only positive frequencies are shown (n ≥ 0). The coefficient c_n is symmetrical about n = 0.

Many mathematical texts provide the Fourier series for an odd function with zero average value (Figure 2-4). If the pulse originates at the origin, then the Fourier series contains only odd harmonics ($a_n = 0$):

$$c_n = V\tau \sin\left(\frac{n\pi}{2}\right) sinc\left(\frac{n\tau}{T}\right) . \tag{2-20}$$

Both the even and odd representations contain sinc. The odd function is further modified by $\sin(n\pi/2)$. When $\tau = T/2$

$$c_n = \frac{VT}{n\pi}\sin^2\left(\frac{n\pi}{2}\right) , \tag{2-21}$$

where $\sin^2(n\pi/2)$ is equal to one when n is odd and zero when n is even.

36 SAMPLING, ALIASING, and DATA FIDELITY

Thus,

$$c_n = \left(\frac{VT}{2}\right)\frac{2}{\pi n} \quad \text{where } n = -3, -1, 1, 3, \cdots, \tag{2-22}$$

This provides the same (absolute) values as Equation 2-18. Figure 2-5 illustrates the amplitude spectrum that provides the same (absolute) values as Figure 2-3. Note that when n is negative, c_n is also negative for odd functions.

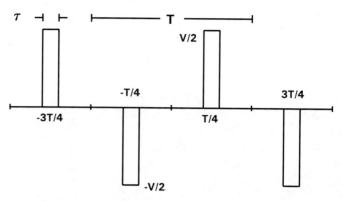

Figure 2-4. An odd function representation: f(-t) = -f(t). The average value (DC component) is zero.

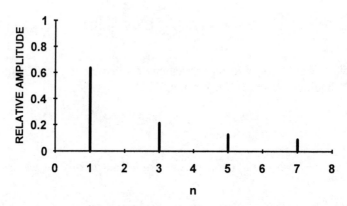

Figure 2-5. Normalized amplitude spectrum of an odd square wave when $\tau = T/2$. This spectrum is seen in many mathematical text books. Only positive frequencies are shown (f ≥ 0). The amplitude has been normalized to VT/2.

FOURIER TRANSFORM 37

The Fourier series and its associated amplitude spectrum indicate that the square wave consists of an infinite number of frequencies. Figure 2-6 illustrates the addition of the first few terms. Figure 2-6a illustrates the DC component and first harmonic (n = 1). The first harmonic is the fundamental frequency of the square wave. Note also the addition of these two terms initially provides a waveform that is greater in amplitude than the square wave. In Figure 2-6b, the 1st, 3rd, and 5th harmonics are added. The result approximates the square wave. Figure 2-6c illustrates the summation up to the 9th harmonic. Clearly, as more terms are added, the summation looks more like the square wave. However, there is oscillation at the edges. The frequency of the oscillation is equal to the frequency of the last term included in the summation. This is the Gibbs phenomenon. Any truncation in the frequency domain (limited number of terms) will create an oscillation or ringing in the time domain.

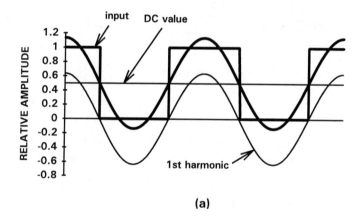

(a)

Figure 2-6. Summation of harmonics. As the number of summed terms increases, the resultant waveform better approximates the original waveform. Using only a finite number will create ringing at sharp transitions (Gibbs phenomenon): (a) DC term and first harmonic (n = 1). Continued next page.

38 SAMPLING, ALIASING, and DATA FIDELITY

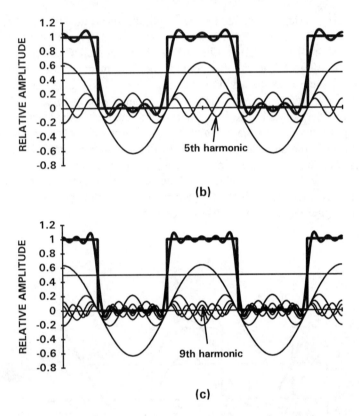

Figure 2-6 (continued). (b) DC term plus 1st, 3rd, and 5th harmonics (n = 1, 3, and 5) and (c) summation with n = 1, 3, 5, 7, and 9.

2.2. FOURIER INTEGRAL

The extension of the Fourier series to aperiodic functions creates the continuous Fourier integral. As T approaches infinity, all pulses illustrated in Figure 2-1 move out beyond the bound leaving only one centered at the origin. This creates a single pulse of amplitude V and width τ. In the frequency plot, f_n approaches the continuous variable f as T approaches infinity.

$$\Delta f = f_{n+1} - f_n = \frac{1}{T} \to 0 \ . \tag{2-23}$$

FOURIER TRANSFORM

The frequency components shown in Figure 2-2 move together (separation approaches zero) and the spectrum becomes continuous (Figure 2-7). Thus,

$$F(f) = \lim_{T \to \infty} c_n = V\tau \, sinc(\tau f) \,. \tag{2-24}$$

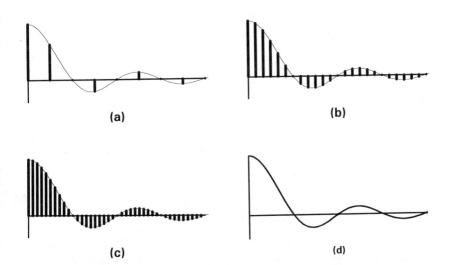

Figure 2-7. Progression of discrete series into a continuous function (a) $\tau/T = 1/2$, (b) $\tau/T = 1/6$, (c) $\tau/T = 1/10$, and (d) continuous function ($\tau/T \to 0$).

In place of the Fourier series we have the forward Fourier transform,

$$F(f) = \int_{-\infty}^{+\infty} f(t) e^{-j2\pi ft} \, dt \,. \tag{2-25}$$

This is the same functional form as Equation 2-9 (page 32) with the limits of integration extended to infinity. The inverse transform of F(f) is called the Fourier integral:

$$f(t) = \int_{-\infty}^{+\infty} F(f) e^{j2\pi ft} \, df \,. \tag{2-26}$$

40 SAMPLING, ALIASING, and DATA FIDELITY

It is the continuous representation of Equation 2-12 (page 32). The transform F(f) is generally complex:

$$F(f) = \Re(f) + j\Im(f) , \qquad (2\text{-}27)$$

where $\Re(f)$ is the real part and $\Im(f)$ is the imaginary part. When expressed in polar coordinates

$$F(f) = |F(f)|e^{j\theta(f)} , \qquad (2\text{-}28)$$

where $|F(f)|$ is the magnitude of the complex F(f) and $\theta(f)$ is the phase. As with the series expansion,

$$|F(f)| = \sqrt{\Re^2(f) + \Im^2(f)} , \qquad (2\text{-}29)$$

and

$$\theta(f) = \tan^{-1}\left(\frac{\Im(f)}{\Re(f)}\right) . \qquad (2\text{-}30)$$

These operations are illustrated in Figure 2-8. The square of the magnitude, $|F(f)|^2$, represents power.

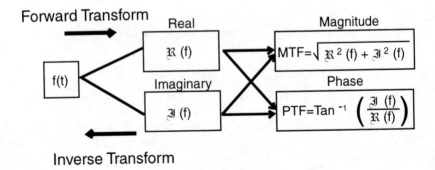

Figure 2-8. Relationship between forward and inverse transforms.

For the two-dimensional case, the transform pairs are

$$F(u,v) = \int_{-\infty}^{+\infty}\int_{-\infty}^{+\infty} f(x,y) e^{-j2\pi(ux+vy)} \, dx\, dy \, , \tag{2-31}$$

$$f(x,y) = \int_{-\infty}^{+\infty}\int_{-\infty}^{+\infty} F(u,v) e^{j2\pi(ux+vy)} \, du\, dv \, , \tag{2-32}$$

and

$$F(u,v) = \Re(u,v) + j\Im(u,v) \, . \tag{2-33}$$

When expressed in polar coordinates

$$F(u,v) = |F(u,v)| e^{j\theta(u,v)} \, . \tag{2-34}$$

2.3. TRANSFORM PROPERTIES

There are three important properties that are frequently used in systems analysis: linearity, translation, and scaling. Linearity means that the transform of the sum of two functions is equal to the sum of the individual transforms. Using double arrows to represent a transform pair, in one dimension,

$$a f_1(t) + b f_2(t) \leftrightarrow a F_1(f) + b F_2(f) \, . \tag{2-35}$$

Translation or shift of the input will change the phase of the transformed function

$$f(t - t_d) \leftrightarrow F(f) e^{-j2\pi t_d f} \, . \tag{2-36}$$

42 SAMPLING, ALIASING, and DATA FIDELITY

With conservation of energy,

$$f\left(\frac{t}{a}\right) \leftrightarrow |a| F(af) . \tag{2-37}$$

Increasing the width of the input function will increase the amplitude and decrease the width of the transformed function. The extension of these properties to two dimensions is straightforward.

Electro-optical imaging systems have several recurring functions that have been assigned a shorthand notation.[1] They are rect(x), sinc(x), and comb(x). Table 2-1 provides the transform pairs.

$$rect\left(\frac{x}{a}\right) = 1 \quad \left|\frac{x}{a}\right| \leq 0.5 \tag{2-38}$$
$$= 0 \quad elsewhere ,$$

$$sinc(ax) = \frac{\sin(\pi a x)}{\pi a x} , \tag{2-39}$$

$$comb(x) = \sum_{-\infty}^{\infty} \delta(x - n\Delta x) , \tag{2-40}$$

where δ(x) is the Dirac delta or impulse function. It has unit value when x = 0 and is zero elsewhere. The function δ(x - nΔx) is an infinite series of impulse functions that only exist at the discrete locations of nΔx. Equivalently, it is an infinite series of impulse functions separated by Δx.

Table 2-1
SOME COMMON TRANSFORM PAIRS

Continuous function	Transform
rect(x/a)	|a| sinc(au)
sinc(x/a)	|a| rect(au)
comb(x/a)	|a| comb(au)

2.4. DISCRETE FOURIER TRANSFORM

The previous equations assumed continuity from $-\infty$ to $+\infty$. In real applications, the signal is sampled for a finite time or distance (Figure 2-9). Figure 2-9d provides the actual data available for analysis. The discrete Fourier transform is used when a finite data set exists. It is an approximation to the continuous transform. The DFT is equivalent to curve fitting the data to a sum of complex sinusoids at discrete frequencies. Amplitudes and phases are adjusted to fit the discrete data points. The number of samples, N, collected is limited by the digital storage capacity, acceptable processing time, or by the number of detectors.

Fundamental to the DFT is that the waveform is periodic (Figure 2-10). If the waveform is divided into N samples (labeled as 0, 1, $\cdots$, N-1), then it is assumed that the same waveform exists from N to 2N-1, 2N to 3N-1, etc.

The continuous integral must be reduced to a finite summation. The signal is sampled every Δt sec, and a total of N samples are acquired.

$$f(k\Delta t) = \sum_{n=0}^{N-1} F(n\Delta f) e^{j\frac{2\pi(k\Delta t)(n\Delta f)}{N}}, \qquad (2\text{-}41)$$

where k is the output index. For each frequency component (k = 0, 1, 2, $\cdots$), the summation is evaluated over all n. The record length is N samples or $N\Delta t$ sec. The discrete transform creates N frequency components each separated by Δf. The frequency resolution Δf is related to the sample time by

$$\Delta f = \frac{1}{N\Delta t}. \qquad (2\text{-}42)$$

As the number of samples increases, the frequency resolution increases (i.e., Δf becomes smaller). As $N \to \infty$, $\Delta f \to 0$ and the transform becomes continuous.

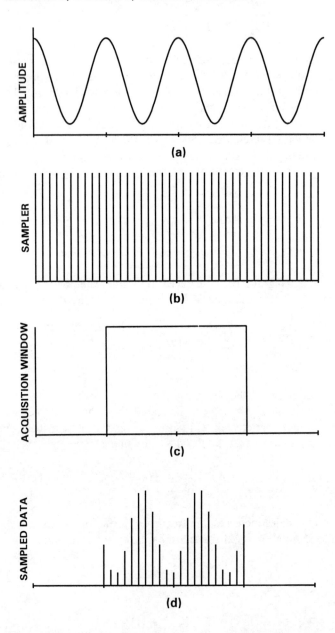

Figure 2-9. (a) Continuous analog signal, (b) continuous sampling train, (c) rectangular acquisition window, and (d) data available for analysis.

FOURIER TRANSFORM 45

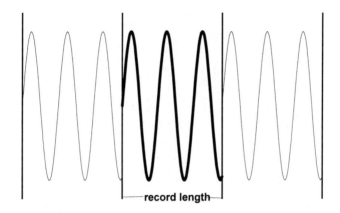

Figure 2-10. The data set is considered continuous. The heavy line represents the acquired data and the light line is the assumed periodicity. Note the discontinuities that exist at the beginning and end of the record.

The Nyquist frequency is

$$f_N = \frac{1}{2\Delta t}, \qquad (2\text{-}43)$$

where f_N is the maximum frequency that the DFT can create. As the sampling time decreases (equivalently the sampling frequency increases), the Nyquist frequency increases. That is, as the sampling rate increases, the highest frequency that can be obtained through the DFT increases.

The inverse discrete transform is

$$F(k\Delta f) = \frac{1}{N} \sum_{n=0}^{N-1} f(n\Delta t) e^{-j\frac{2\pi(n\Delta t)(k\Delta f)}{N}}. \qquad (2\text{-}44)$$

The above equations include the scaling that exists between Δf and Δt. A computer algorithm operates on an array that is identified by indices only.

46 SAMPLING, ALIASING, and DATA FIDELITY

Scaling factors are added only when displaying the data. Using only indices,

$$X(k) = \frac{1}{N} \sum_{n=0}^{N-1} x(n) e^{-j\frac{2\pi nk}{N}}, \qquad (2\text{-}45)$$

where
    x(n) is the value of data point n. It is the value of f(nΔt)
    X(k) is the value of the transformed data. It is the value of F(kΔf)
    n is the input data set index that varies from 0 to N-1
    k is the output data set index that varies from 0 to N-1.

The inverse transform is

$$x(k) = \sum_{n=0}^{N-1} X(n) e^{j\frac{2\pi nk}{N}}, \qquad (2\text{-}46)$$

where n has units of samples and k has units of cycles per sample. When displayed, the graphs are often in units of nΔt and kΔf.

In terms of the indices, the Nyquist frequency occurs at N/2. In terms of normalized units, n/N, the sampling frequency is 1 and the Nyquist frequency is 0.5. The DFT is periodic with period N so that

$$X(k) = X(k+N). \qquad (2\text{-}47)$$

When f(x) is real, the magnitude of the transform from N/2 + 1 to (N - 1) is a mirror image of the magnitude from 1 to (N/2 - 1). That is, the data are symmetrical about N/2. Figure 2-11a illustrates the computed DFT when N = 16. The Nyquist frequency occurs at N = 8. In Figure 2-11b, the data have been centered with the array indices listed. For electrical engineering applications (time domain analysis) only the data points from 0 to N/2 are plotted. N/2 corresponds to 1/2T Hz. The increments are $\Delta f = 1/(2\,T\,N)$.

FOURIER TRANSFORM 47

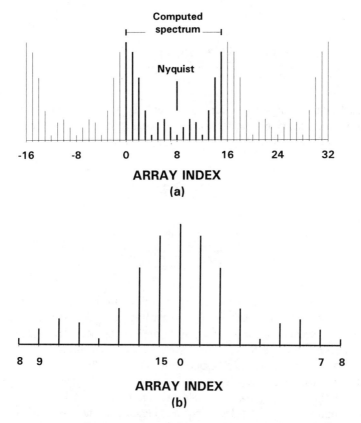

Figure 2-11. DFT when N = 16. (a) The heavy lines indicate the computed DFT and the light lines represent the assumed periodicity. (b) Centered data with elements 8 though 15 mirrored about zero. Each increment on the x-axis scale is equal to $1/(N\Delta t)$.

The extension to two dimensions is straightforward. For an array that contains N × M elements, the two-dimensional DFT transform pairs are

$$F(u,v) = \frac{1}{NM} \sum_{x=0}^{N-1} \sum_{y=0}^{M-1} f(x,y) e^{-j2\pi\left(\frac{ux}{N} + \frac{vy}{M}\right)} \quad (2\text{-}48)$$

48 SAMPLING, ALIASING, and DATA FIDELITY

and

$$f(x,y) = \sum_{u=0}^{N-1} \sum_{v=0}^{M-1} F(u,v) e^{j2\pi\left(\frac{ux}{N} + \frac{vy}{M}\right)}. \qquad (2\text{-}49)$$

The signal is sampled at intervals of Δx and Δy in the x and y directions. Starting at an arbitrary origin, f(x,y) is evaluated at f(nΔx,mΔy) for n = 0, 1, $\cdots$, N-1 and m = 0, 1, $\cdots$, M-1. The sampling intervals are

$$\Delta u = \frac{1}{N \Delta x} \qquad (2\text{-}50)$$

and

$$\Delta v = \frac{1}{M \Delta y} \qquad (2\text{-}51)$$

where N and M are datels and can be any value. In practice, several values occur more frequently than others. When creating computer generated scenes, the number of scenels tends to be a power of 2 (64 × 64, 128 × 128, 256 × 256, etc.). For staring arrays, N and M are usually equal to the number of horizontal and vertical pixels, respectively. Table 2-2 lists some common detector array sizes. However, for c/d/c/d/c systems (See Figure 1-17, page 24), the frame grabber can digitize the analog signal at any rate. If there are more datels than pixels, the Nyquist frequency associated with the DFT is higher than the detector array Nyquist frequency.

Table 2-2
COMMON DETECTOR ARRAY SIZES

DESCRIPTION	N (Horizontal)	M (Vertical)
Monochrome consumer camera	640	480
Color consumer camera (NTSC)	768	494
Color consumer camera (PAL)	752	582
Scientific arrays	512 or 1024	512 or 1024
High definition TV	720	1280

FOURIER TRANSFORM 49

The two-dimensional transform creates mirror images about N/2 and M/2. The quadrants must be swapped[2] to center the data (Figure 2-12). After centering, the graph axes are -(N/2)-1 to N/2 and -(M/2)-1 to M/2 cycles where N and M are the (even) number of datels associated with the image.

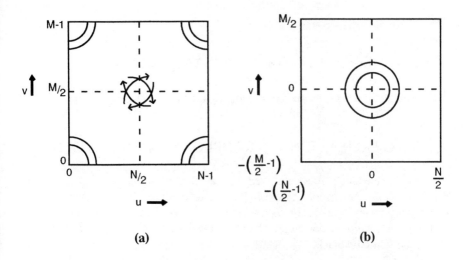

Figure 2-12. Quadrant swapping to place the origin in the center of the figure. (a) Transformed data before swapping. The arrows indicate the swapping. (b) Centered data. Centering data usually simplifies analysis and interpretation but is not necessary because all the information is present in the "unswapped" representation.

Different frequency domains are discussed in Section 9.1., *Frequency Domain*. Given the horizontal and vertical fields-of-view, the axes in object space are

$$u\text{-}axes: \quad -\left(\frac{N}{2}-1\right)\frac{1}{HFOV} \quad to \quad \frac{N}{2 \cdot HFOV} \quad \frac{cycles}{mrad} \quad (2\text{-}52)$$

and

$$v\text{-}axes: \quad -\left(\frac{M}{2}-1\right)\frac{1}{VFOV} \quad to \quad \frac{M}{2 \cdot VFOV} \quad \frac{cycles}{mrad}. \quad (2\text{-}53)$$

50 SAMPLING, ALIASING, and DATA FIDELITY

The frequency increments are $\Delta u = 1/\text{HFOV}$ and $\Delta v = 1/\text{VFOV}$. If the total array size is $(d_{ARRAY-H}, d_{ARRAY-V})$ then the axes in image space are

$$u\text{-axes:} \quad -\left(\frac{N}{2} - 1\right)\frac{1}{d_{ARRAY-H}} \quad \text{to} \quad \frac{N}{2 \cdot d_{ARRAY-H}} \quad \frac{cycles}{mm} \qquad (2\text{-}54)$$

and

$$v\text{-axes:} \quad -\left(\frac{M}{2} - 1\right)\frac{1}{d_{ARRAY-V}} \quad \text{to} \quad \frac{M}{2 \cdot d_{ARRAY-V}} \quad \frac{cycles}{mm} \; . \qquad (2\text{-}55)$$

The frequency increments are $\Delta u = 1/d_{ARRAY-H}$ and $\Delta v = 1/d_{ARRAY-V}$.

2.5. TWO-DIMENSIONAL TRANSFORM REPRESENTATION

Two-dimensional functions require a three-dimensional graph (two dimensions plus amplitude), and this is more difficult to illustrate. Assignment of intensity levels is arbitrary. An object may be white on a black background or black on a white background. The choice is simply whether the background should be white or black.

Figure 2-13a illustrates a horizontal sine wave pattern that has no variation in the vertical direction. The Fourier transform provides both the positive and negative frequencies as shown in Figure 2-13c. This is identical to the one-dimensional transforms provided earlier. Here distance replaces time and F(u) replaces F(f). Figure 2-13d shows the typical output from a DFT.

Figure 2-14 illustrates a vertical sine wave with the frequency component on the v-axis. Similarly, a vertical square wave (Figure 2-15) contains frequency components only on the v-axis.

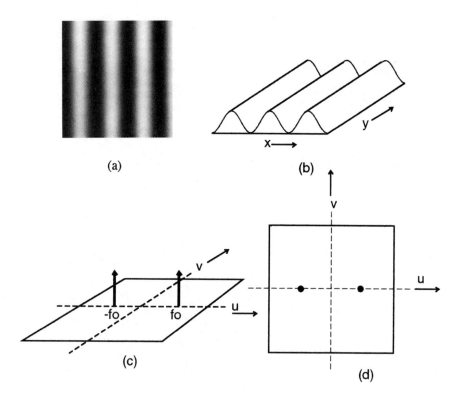

Figure 2-13. Horizontal cosine wave. (a) Gray levels represent intensity,[3] (b) three-dimensional contour plot, (c) three-dimensional representation of the frequency components, and (d) gray level representation of the two-dimensional frequency plane.

52 SAMPLING, ALIASING, and DATA FIDELITY

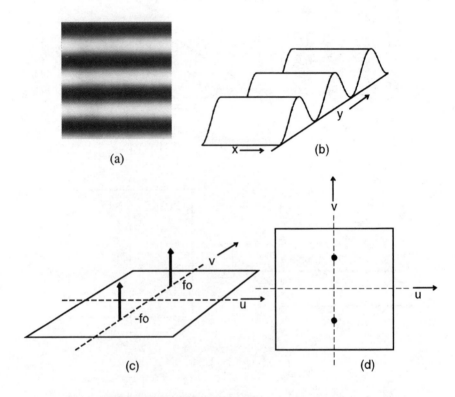

Figure 2-14. Vertical cosine wave. (a) Gray levels represent intensity,[3] (b) three-dimensional contour plot, (c) three-dimensional representation of the frequency components, and (d) gray level representation of the two-dimensional frequency plane.

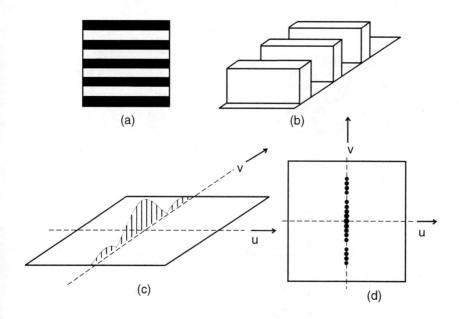

Figure 2-15. Vertical square wave. The sinc envelope is shown. See Figure 2-13 for the description of (a), (b), (c), and (d).

Figure 2-16 illustrates a square wave that has been rotated by θ_o. The rotational property states that if an image is rotated by θ_o, then the spectrum is also rotated by θ_o:

$$f(r, \theta - \theta_o) \leftrightarrow F(\rho, \phi - \theta_o) . \qquad (2\text{-}56)$$

The square wave period is composed of x and y components:

$$d_o = \sqrt{d_x^2 + d_y^2} . \qquad (2\text{-}57)$$

Similarly, the frequency components are also rotated by θ. In polar coordinates,

$$|F(r,\theta)| = \sqrt{F^2(u) + F^2(v)} . \qquad (2\text{-}58)$$

A function that is separable in cartesian coordinates may not be separable in polar coordinates and vice versa.

54 SAMPLING, ALIASING, and DATA FIDELITY

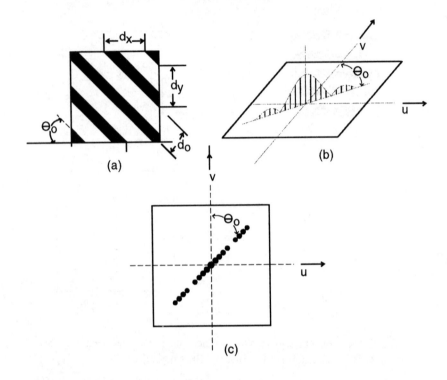

Figure 2-16. Square wave (a) rotated by θ_o. (b) Three-dimensional representation of the frequency components, and (c) gray level representation of the two-dimensional frequency plane. Because the intensity variation is in one direction only, the frequency components lie on a line. The angle in frequency space is the same as that in object space.

As with the one-dimensional DFT, the two-dimensional DFT is based on the assumption that the image is repetitive (Figure 2-17). Figure 2-18 illustrates a triangle and its DFT. Figures 2-19 and 2-20 illustrate the DFT of a square and rectangle respectively. In both directions, the amplitude envelope follows a sinc function.

FOURIER TRANSFORM

The first zero of the rectangular pulse spectrum, sinc(τu), occurs at $1/\tau$. If the image consists of N_{SAMPLE} samples, each datel represents $1/N_{SAMPLE}$ of the image. Given the HFOV, each datel is HFOV/N_{SAMPLE} mrad. If the pulse width is N_o datels, then it is N_o/N_{SAMPLE} of the image width or ($N_o \cdot$ HFOV)/N_{SAMPLE} mrad. The first zero ($1/\tau$) occurs at N_{SAMPLE}/N_o cycles or $N_{SAMPLE}/(N_o \cdot$ HFOV) cycles/mrad. The maximum frequency is $N_{SAMPLE}/2$ cycles or $N_{SAMPLE}/(2 \cdot$ HFOV) cycles/mrad. When expressed as a fractional part of f_{MAX}, the first zero will occur at

$$Fraction = \frac{\left(\frac{1}{\tau}\right)}{f_{MAX}} = \frac{\frac{N_{SAMPLE}}{N_o \cdot HFOV}}{\frac{N_{SAMPLE}}{2 \cdot HFOV}} = \frac{2}{N_o}. \qquad (2\text{-}59)$$

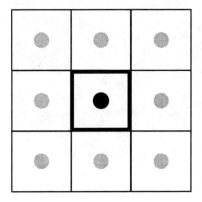

Figure 2-17. Only data available within the center square exist. The two-dimensional DFT is based on the assumption that the signal is repetitive in two dimensions.

56 SAMPLING, ALIASING, and DATA FIDELITY

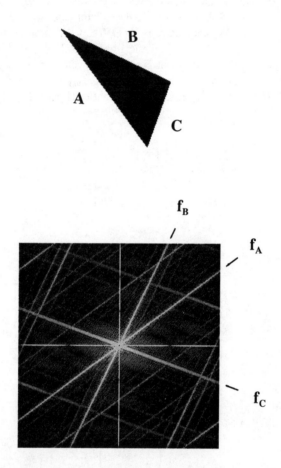

Figure 2-18. The DFT returns the frequency components of the image. The dominant components are related to the dominant image features. With an edge, the frequency components are normal to the edge (see Figure 2-16). Edges A, B, and C produce f_A, f_B, and f_C, respectively. The gray scale is logarithmic.

FOURIER TRANSFORM 57

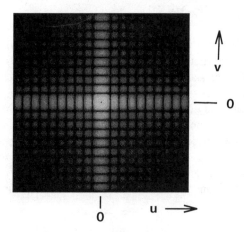

Figure 2-19. DFT of a square. The amplitude spectrum is F(u,v) = sinc(τu)sinc(τv). The square is 20 × 20 datels. The first zero occurs at 1/10 of the distance from the center to the maximum frequency. The intensity along the u and v axes is magnitude of the sinc function. F(u,v) is small when u > 1/τ or v > 1/τ. The logarithmic gray scale emphasizes high frequency response.

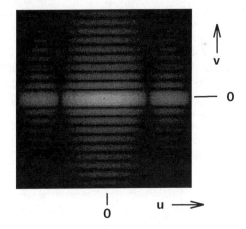

Figure 2-20. DFT of a rectangle 4 × 20 datels. The amplitude spectrum is F(u,v) = sinc(τu/5)sinc(τv). The first horizontal zero occurs at 1/2 of the distance from the center to the maximum frequency. The first vertical zero occurs at 1/10 of the distance. F(u,v) is small when u > 5/τ or v > 1/τ. The logarithmic gray scale emphasizes high frequency response.

58 SAMPLING, ALIASING, and DATA FIDELITY

2.6. WINDOWS

As stated at the beginning of Section 2.4, the DFT is equivalent to curve fitting the data to a sum of complex sinusoids at discrete frequencies. The amplitudes and phases are adjusted to fit the discrete data points. If the number of sinusoids selected is equal to the data length, the data can be fitted exactly and uniquely to the DFT frequencies. The DFT frequencies are constrained to be integer multiples of the sampling frequency divided by the record length. When a frequency component in a signal does not exactly fit onto a bin (i.e., when n = 0, 1, ···), the DFT algorithm must assign a nonzero amplitude to many unrelated sinusoids to fit the data exactly. This effect is called leakage. Leakage can be a serious problem when analyzing periodic or almost periodic signals.

Figure 2-9 (page 44) illustrated how to create a finite data set using a window, w(t). Mathematically,

$$f_{DATA}(t) = f_{INFINITE}(t) \cdot w(t) . \qquad (2\text{-}60)$$

Because multiplication in the time domain becomes a convolution in the frequency domain (further discussed in Section 3.1., *Linear System Theory*)

$$F_{DATA}(f) = F_{INFINITE}(f) * W(f) . \qquad (2\text{-}61)$$

For a sinusoid, $F_{INFINITE}(f) = f_o$. With a rectangular window, $W(f) = \text{sinc}(fT_{WINDOW})$ where T_{WINDOW} is the window width and $T_{WINDOW} = N\Delta t$. Rather than a single frequency at f_o, convolution creates a spectrum that becomes $\text{sinc}(T_{WINDOW}(f - f_o))$. Let the sinusoid period, t_o, be $k\Delta t$ sec long. The discrete transform provides

$$F_{DATA}(f_n) = \text{sinc}\left(T_{WINDOW}\left(\frac{n}{N\Delta t} - \frac{1}{k\Delta t}\right)\right) = \text{sinc}\left(n - \frac{N}{k}\right) . \qquad (2\text{-}62)$$

The number of cycles captured by the window depends on the window width. Let T_{WINDOW} enclose an integer number of cycles. Then N/k is an integer and sinc(n-N/k) is zero for all n unless n = N/k. If T_{WINDOW} does not enclose an integral number of cycles, then sinc(n-T_{WINDOW}/k) has finite values at many other frequencies (Figure 2-21). This is leakage. In the frequency domain, as the window width increases, the width of sinc(fT_{WINDOW}) decreases (scaling property, Equation 2-37, page 42).

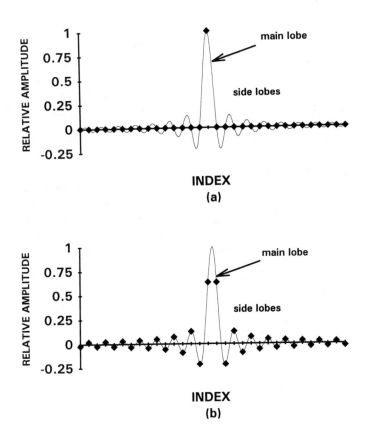

Figure 2-21. The light line is the amplitude spectrum of a finite length sinusoid. The dots are the discrete values created by the DFT. (a) Integer number of cycles in the window and (b) nonintegral number.

60 SAMPLING, ALIASING, and DATA FIDELITY

The average value was omitted from the theoretical amplitude spectrum shown in the previous sections. However, the average value exists in the computed DFT. The DFT provides both positive and negative frequencies with the energy split equally between the two. Only one DC value is computed so that it will appear to have twice the energy compared to the other frequencies. The measured average value also depends on the fractional number of cycles captured. As a result there is some ambiguity in the computed DC value. Usually the DC component is ignored (the DFT is used to verify frequency components, not average values). It can be removed from $f_{DATA}(t)$ prior to performing the DFT.

The rect(t) window exists due to the finiteness of the data set. By multiplying by another window that tapers the data amplitude at the beginning and end of the record, leakage effects can be reduced. When using an additional window, be sure that the most interesting parts of the signal are in the center of the window. Otherwise, the tapering may cause significant errors. That is, windows are effective when the data set is large. Any DC offset should be removed before using a window.

Various windows are evaluated until the desired data features are emphasized. For example, a window can significantly reduce frequency amplitudes far from f_o (Figure 2-21b). However, the window will also reduce amplitudes near f_o. The various windows trade main lobe width (that centered on f_o) with leakage at frequencies far from f_o. Two popular windows are the Hamming and Hanning windows. In terms of data indices, the Hanning or raised cosine is

$$w(n) = \frac{1}{2}\left(1 - \cos\left(\frac{2\pi n}{N}\right)\right) \quad \text{where } n = 0, 1, \cdots, N-1 . \quad (2\text{-}63)$$

The Hamming window is

$$w(n) = 0.5435 + 0.4565\left(1 - \cos\left(\frac{2\pi n}{N}\right)\right) \quad (2\text{-}64)$$

$$\text{where } n = 0, 1, \cdots, N-1 .$$

FOURIER TRANSFORM   61

The discrete data set becomes

$$x_{DATA}(n) = x(n) \cdot w(n) \ . \tag{2-65}$$

2.7. TYPICAL TEST PATTERNS

Numerous targets can be created by multiplying an infinite series by an appropriate window width. Electronic imaging system test patterns typically consist of three or four bars. They may be considered as an infinite square wave multiplied by a window that just passes three or four bars. In one-dimensional space (Figure 2-22), the three-bar pattern is

$$f_{3-BAR}(x) = f_{SERIES}(x) \cdot w(x) \ . \tag{2-66}$$

Because multiplication in the space domain becomes a convolution in the frequency domain,

$$F_{3-BAR}(u) = F_{SERIES}(u) * W(u) \ . \tag{2-67}$$

When the bar width is $T/2$, the window that passes three bars has a width of $5T/2$. Its frequency response is

$$W(u) = sinc\left(\frac{5Tu}{2}\right) \ . \tag{2-68}$$

This is convolved with the series given by Equation 2-13 (page 33).

When only a few bars are used, it is easier to determine the frequency components using the shift principle (Equation 2-36, page 41). Each bar has a sinc transform and three bars provide three separate sinc functions. If the center pulse is at the origin, then the other pulses exist at -T and T:

$$F_{3-BAR}(u) = \frac{VT}{2} sinc\left(\frac{Tu}{2}\right) \left(1 + e^{j2\pi Tu} + e^{-j2\pi Tu}\right) \tag{2-69}$$

or

$$F_{3-BAR}(u) = \frac{VT}{2} sinc\left(\frac{Tu}{2}\right) [1 + 2\cos(2\pi Tu)] \ . \tag{2-70}$$

62 SAMPLING, ALIASING, and DATA FIDELITY

Note that $F_{3\text{-BAR}}(u)$ has three times more energy than a single bar. Because the frequency components are used for image analysis, the amplitude is normalized to unity. Manipulating the expression provides[4] (Figure 2-23a)

$$F_{3\text{-BAR}}(u) = \frac{VT}{2}\frac{T}{\pi u}\frac{\sin\left(\dfrac{6\pi Tu}{2}\right)}{\cos\left(\dfrac{\pi Tu}{2}\right)}. \qquad (2\text{-}71)$$

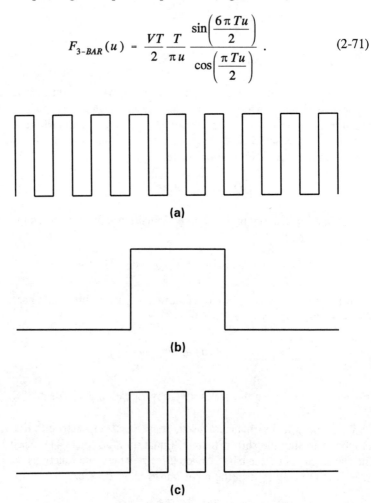

Figure 2-22. A three-bar pattern can be generated by multiplying an infinite series with a window. (a) Infinite series, (b) window, and (c) three-bar target remaining after the window. It consists of three bars and two spaces.

Similarly, a four-bar target amplitude spectrum can be created by the shift property. Let the pulses reside at -3T/2, -T/2, T/2, and 3T/2. Then

$$F_{4\text{-}BAR}(u) = \frac{VT}{2} \text{sinc}\left(\frac{Tu}{2}\right)\left(e^{-j3\pi Tu} + e^{-j\pi Tu} + e^{j\pi Tu} + e^{j3\pi Tu}\right) \quad (2\text{-}72)$$

or

$$F_{4\text{-}BAR}(u) = \frac{VT}{2} \text{sinc}\left(\frac{Tu}{2}\right)[2\cos(\pi Tu) + 2\cos(3\pi Tu)], \quad (2\text{-}73)$$

where $F_{4\text{-}BAR}(u)$ has four times more energy than a single bar. As with the three-bar target amplitude spectrum, the amplitude is typically normalized to unity for image analysis (Figure 2-23b). As the number of bars increases, the width of the spectra about the square wave harmonics decreases. In the limit as the window width approaches infinity, the frequency components narrow and Figure 2-3 (page 35) is obtained. Figure 2-24 illustrates the DFT of a three-bar target.

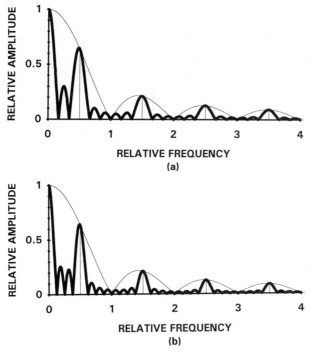

Figure 2-23. Magnitude of the (a) three-bar target amplitude spectrum and (b) four-bar target amplitude spectrum.

64 SAMPLING, ALIASING, and DATA FIDELITY

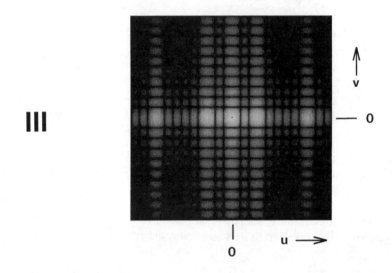

Figure 2-24. DFT of a three-bar pattern. Each bar is 4 × 20 pixels (see Figure 2-20, page 57). The function $F_{3\text{-BAR}}(u,v)$ is $F_{3\text{-BAR}}(u)\text{sinc}(\tau v)$, where $F_{3\text{-BAR}}(u)$ is given by Equation 2-70 and illustrated in Figure 2-23a. The first vertical zero occurs at 1/10 of the distance from the center to the maximum frequency. The function F(v) is identical to that illustrated in Figures 2-19 and 2-20. The gray scale is logarithmic.

2.8. REFERENCES

1. J. D. Gaskill, *Linear Systems, Fourier Transforms, and Optics*, John Wiley and Sons, New York (1978).
2. A. R. Weeks, Jr., *Fundamentals of Electronic Image Processing*, page 156, SPIE Optical Engineering Press, Bellingham, WA (1996).
3. Photographic sinusoids provided by Sine Patterns, 236 Henderson Drive, Penfield, NY 14526.
4. C. S. Williams and O. A. Becklund, *Introduction to the Optical Transfer Function*, pp. 30-31, John Wiley & Sons, New York (1989).

3
LINEAR SYSTEM THEORY

Linear system theory was developed for electronic circuitry and has been extended to optical, electro-optical, and mechanical systems. It forms an indispensable part of system analysis. For modeling purposes, electronic imaging systems are characterized as linear spatial-temporal systems that are shift-invariant with respect to both time and two spatial dimensions. Although space is three-dimensional, an imaging system displays only two dimensions.

Electrical filters are different from spatial filters in two ways. They are single-sided in time and must satisfy the causality requirement that no change in the output may occur before the application of an input. Optical filters are double-sided in space. Electrical signals may be either positive or negative, whereas optical intensities are always positive. As a result, optical designers and circuit designers often use different terminology.

With a linear-shift-invariant (LSI) system, signal processing is linear. A shift of the input causes a corresponding shift in the output. For electronic circuits, the shift is in time, whereas for optical systems, the shift is in space. In addition, the input-to-output mapping is single-valued. For optical systems, one addition condition exists: the radiation must be incoherent if the system is to be described in terms of radiance (e.g., units of watt/m^2).

A sampled data system may be considered "globally" shift invariant on a macro-scale. For example, with an electronic imaging system, as a target moves from the top of the field-of-view to the bottom, the image also moves from the top to the bottom. On a micro-scale, moving a point source across a single detector does not change the detector output. That is, the system is not shift-invariant on a micro-scale. Similarly, with electronic circuits, large time shifts of the input produce corresponding large time shifts in output. A shift during a sample time, does not change the output. Here also, the system is not shift-invariant on a micro-scale.

Single-valued mapping only occurs with nonnoisy and nonquantized systems. No system is truly noiseless but can be approximated as one when the signal has sufficient amplitude. With digital systems, the output is quantized. The smallest output is the least significant bit and the analog-to-digital converter generally limits the largest signal. If the signal level is large compared to the LSB, then

66 SAMPLING, ALIASING, and DATA FIDELITY

the system can be treated as quasi-linear over a restricted region.

In spite of the disclaimers mentioned, systems are treated as quasi-linear and quasi-shift-invariant over a restricted operating region to take advantage of the wealth of mathematical tools available. An LSI system merely modifies the amplitude and phase of the input. These are specified by the modulation transfer function and phase transfer function.

The symbols used in this book are summarized in the *Symbol List* (page xiii) which appears after the *Table of Contents*.

3.1. LINEAR SYSTEM THEORY

A linear system provides single mapping from an input to an output. Superposition allows the addition of individual inputs to create an unique output. Let h{ } be a linear operator that maps one function, f(t), into another function, g(t):

$$h\{f(t)\} = g(t) . \qquad (3\text{-}1)$$

Let the response to two inputs, $f_1(t)$ and $f_2(t)$, be $g_1(t)$ and $g_2(t)$:

$$h\{f_1(t)\} = g_1(t) \quad \text{and} \quad h\{f_2(t)\} = g_2(t) . \qquad (3\text{-}2)$$

For a linear system, the response to a sum of inputs is equal to the sum of responses to each input acting separately. For any arbitrary scale factors, the superposition principle states

$$h\{a_1 f_1(t) + a_2 f_2(t)\} = a_1 g_1(t) + a_2 g_2(t) . \qquad (3\text{-}3)$$

When passing the signal through another linear system, the new operator provides:

$$h_2\{g(t)\} = h_2\{h_1\{f(t)\}\} . \qquad (3\text{-}4)$$

3.1.1. TIME-VARYING SIGNALS

An electrical signal can be thought of as the sum of an infinite array of impulses (Dirac delta functions) located inside the signal boundaries. Thus, the signal can be decomposed into a series of weighted Dirac delta functions (Figure 3-1):

$$v_{IN}(t) = \sum_{t'=-\infty}^{\infty} v_{IN}(t')\delta(t-t')\Delta t', \qquad (3\text{-}5)$$

where $\delta(t-t')$ is the Dirac delta or impulse function at $t = t'$ and $v_{IN}(t')$ is the signal evaluated at t'.

If $v_{IN}(t)$ is applied to a linear circuit, using $h\{\ \}$ will provide the output

$$v_{OUT}(t) = \sum_{t'=-\infty}^{\infty} h\{v_{IN}(t')\delta(t-t')\Delta t'\}. \qquad (3\text{-}6)$$

The wave form $v_{IN}(t')$ can be considered the weightings, a_i, and $\delta(t-t')$ be $f(t)$ in Equation 3-3. That is, the input has been separated into a series of functions $a_1 f_1(t) + a_2 f_2(t) + \cdots$. As $\Delta t' \to 0$, this becomes the convolution integral

$$v_{OUT}(t) = \int_{-\infty}^{\infty} v_{IN}(t') h\{\delta(t-t')\}\, dt', \qquad (3\text{-}7)$$

and is symbolically represented by

$$v_{OUT}(t) = v_{IN}(t) * h(t), \qquad (3\text{-}8)$$

where $*$ indicates the convolution operator (other texts may use $\star$ or $\otimes$). The system impulse response is $h\{\delta(t)\}$. In Figure 3-2, each impulse response is given for each t'. The individual impulse responses are added to produce the output. This addition is the definition of superposition.

68 SAMPLING, ALIASING, and DATA FIDELITY

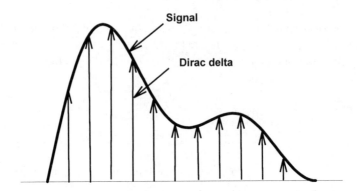

Figure 3-1. A time-varying signal can be decomposed into a series of closely spaced impulses with amplitude equal to the signal at that value. They are shown widely separated for clarity and should not be confused with sampling.

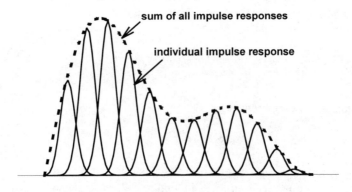

Figure 3-2. The linear operator, h{ }, transforms each input Dirac delta into an output impulse response. The sum of the impulse responses creates the output. Although shown separated for clarity, the impulse responses have infinitesimal separations.

LINEAR SYSTEM THEORY 69

The convolution theorem of Fourier analysis states that the Fourier transform of the convolution of two functions equals the product of the transform of the two functions. That is, multiple convolutions are equivalent to multiplication in the frequency domain. Using frequency domain analysis is convenient because multiplications are easier to perform and easier to visualize than convolutions. Then

$$V_{OUT}(f) = V_{IN}(f) H(f) . \qquad (3\text{-}9)$$

Let the input be an impulse. The Fourier transform of an impulse response is

$$V_{IN}(f) = 1 . \qquad (3\text{-}10)$$

That is, all frequencies are present. Because the input to the system contains all frequencies, the output must be a function of the system only. Any variation in phase or amplitude must be due to the system itself. The output is

$$V_{OUT}(f) = H(f) . \qquad (3\text{-}11)$$

3.1.2. SPATIALLY VARYING SIGNALS

The one-dimensional electronic signal decomposition can be easily extended to two-dimensional imagery. An object can be thought of as the sum of an infinite array of impulses located inside the target boundaries. Thus, an object can be decomposed into a two-dimensional array of weighted Dirac delta functions, $\delta(x - x'), \delta(y - y')$:

$$o(x,y) = \sum_{x'=-\infty}^{\infty} \sum_{y'=-\infty}^{\infty} o(x',y') \delta(x - x') \delta(y - y') \Delta x' \Delta y' . \qquad (3\text{-}12)$$

An optical system with operator $h\{\ \}$ produces an image

$$i(x,y) = \sum_{x'=-\infty}^{\infty} \sum_{y'=-\infty}^{\infty} h\{o(x',y') \delta(x - x') \delta(y - y') \Delta x' \Delta y'\} . \qquad (3\text{-}13)$$

For small increments, this becomes the convolution integral

$$i(x,y) = \int_{-\infty}^{\infty}\int_{-\infty}^{\infty} o(x',y')\, h\{\delta(x-x')\delta(y-y')\}\, dx'\, dy' \qquad (3\text{-}14)$$

and is symbolically represented by

$$i(x,y) = o(x,y) ** h(x,y). \qquad (3\text{-}15)$$

where $**$ represents the two-dimensional convolution. The function $h(x,y)$ is the optical system's response to an input impulse. The resulting image is the point spread function (PSF). For an ideal circular aperture, the central portion of the PSF is the Airy disk from which the Rayleigh criterion and other resolution metrics are derived. In one-dimension $h(x)$ is the line spread function, LSF. The LSF is the resultant image produced by the imaging system when viewing an ideal line. There is no equivalent interpretation for $h(t)$ for electrical circuits. The function $h(t)$ is simply the impulse response.

The transform of the convolution becomes a multiplication in the frequency domain

$$I(u,v) = O(u,v)\, H(u,v). \qquad (3\text{-}16)$$

If all frequencies are present in the image (i.e., an impulse), $O(u,v) = 1$, then $I(u,v) = H(u,v)$.

3.2. THE ELECTRONIC IMAGING SYSTEM

The electronic imaging system response consists of both the optical response and the electric response: $h(x,y,t)$. Time and spatial coordinates are treated separately. For example, optical elements do not generally change with time and therefore are characterized only by spatial coordinates. Similarly, electronic circuitry exhibits only temporal responses. The detector provides the interface between the spatial and temporal components, and its response depends on both temporal and spatial quantities. The conversion of two-dimensional optical information to a one-dimensional electrical response assumes a linear photo-detection process. Implicit in the detector response is the conversion from input photon flux to output voltage.

LINEAR SYSTEM THEORY

The optical transfer function (OTF) plays a key role in the theoretical evaluation and optimization of an optical system. The modulation transfer function (MTF) is the magnitude and the phase transfer function (PTF) is the phase of the complex-valued OTF. In many applications, the OTF is real-valued and positive so that the OTF and MTF are equal. When focus errors or aberrations are present, the OTF may become negative or even complex valued. Electronic circuitry also can be described by an MTF and PTF. The combination of the optical MTF and the electronic MTF creates the electronic imaging system MTF. The MTF is the primary parameter used for system design, analysis, and specifications. When coupled with the three-dimensional noise parameters,[1] the MTF and PTF uniquely define system performance.

Symbolically

$$OTF(u,v) = H_{SPATIAL}(u,v) = MTF(u,v)\, e^{jPTF(u,v)} \qquad (3\text{-}17)$$

and

$$H_{ELECTRONICS}(f) = MTF(f)\, e^{jPTF(f)}. \qquad (3\text{-}18)$$

With appropriate scaling, the electronic frequencies can be converted into spatial frequencies. This is symbolically represented by $f \to u$. The electronic circuitry is assumed to modify the horizontal signal only (although this depends on the system design). The combination of spatial and electronic responses is sometimes called the system OTF:

$$OTF_{SYSTEM}(u,v) = MTF(u,v)\, MTF(f \to u)\, e^{j[PTF(u,v) + PTF(f \to u)]}. \qquad (3\text{-}19)$$

The displayed image can be computed through the Fourier transform pairs. The two-dimensional transform of the object is

$$O(u,v) = \int_{-\infty}^{+\infty}\int_{-\infty}^{+\infty} o(x,y)\, e^{-j2\pi(ux+vy)}\, dx\,dy. \qquad (3\text{-}20)$$

72 SAMPLING, ALIASING, and DATA FIDELITY

This multiplied by the imaging system OTF (Equation 3-19) and the final displayed image is

$$i_D(x,y) = \int_{-\infty}^{+\infty}\int_{-\infty}^{+\infty} O(u,v)\,OTF(u,v)\,e^{j2\pi(ux+vy)}\,du\,dv \; . \qquad (3\text{-}21)$$

Given the system OTF, the response to any arbitrary input can be calculated.

For mathematical convenience, the horizontal and vertical responses are considered separable (usually coincident with the detector array axes). The electrical response is considered to affect only the horizontal response

$$H_{SYSTEM}(u) = H_{SPATIAL}(u)\,H_{ELECTRONICS}(f \rightarrow u) \qquad (3\text{-}22)$$

and

$$H_{SYSTEM}(v) = H_{SPATIAL}(v) \; . \qquad (3\text{-}23)$$

A system is composed of many components that respond to either spatial or temporal signals. For independent MTFs,

$$MTF_{SYSTEM}(u,v) = \prod_{i=1}^{N}\prod_{j=1}^{M} MTF_i(u,v)\,MTF_j(f \rightarrow u) \; . \qquad (3\text{-}24)$$

While individual lens elements each have their own MTF, the MTF of the lens *system* is not usually the product of the individual MTFs. This occurs because one lens may minimize the aberrations created by another.

3.3. MTF and PTF INTERPRETATION

The system MTF and PTF alter the image as it passes through the circuitry. For linear-shift-invariant systems, the PTF simply indicates a spatial or temporal shift with respect to an arbitrarily selected origin. An image where the MTF is drastically altered is still recognizable, whereas large PTF nonlinearities can destroy recognizability. Modest PTF nonlinearity may not be noticed visually except those applications where target geometric properties must be preserved (i.e., mapping or photogrammetry). Generally, PTF nonlinearity increases as the spatial frequency increases. Because the MTF is small at high frequencies, the nonlinear-phase-shift effect is diminished.

The MTF is the ratio of output modulation to input modulation normalized to unity at zero frequency. While the modulation changes with system gain, the MTF does not. The input can be as small as desired (assuming a noiseless system with high gain) or it can be as large as desired because the system is assumed not to saturate.

Modulation is the variation of a sinusoidal signal about its average value (Figure 3-3). It can be considered as the AC amplitude divided by the DC level. The modulation is:

$$MODULATION = M = \frac{V_{MAX} - V_{MIN}}{V_{MAX} + V_{MIN}} = \frac{AC}{DC} . \quad (3\text{-}25)$$

where V_{MAX} and V_{MIN} are the maximum and minimum signal levels, respectively. The modulation transfer function is the output modulation produced by the system divided by the input modulation as a function of frequency:

$$MTF(f) = \frac{M_{OUTPUT}(f)}{M_{INPUT}(f)} . \quad (3\text{-}26)$$

The concept is presented in Figure 3-4. Three input and output signals are plotted in Figures 3-4a and 3-4b, respectively, and the resultant MTF is shown in Figure 3-4c. As a ratio, the MTF is a relative measure with values ranging from zero to one.

The MTF is a measure of how well a system will faithfully reproduce the input. The highest frequency that can be faithfully reproduced is the system cutoff frequency. If the input frequency is above the cutoff, the output will be proportional to the signal's average value with no modulation.

Electronic signals may be both positive and negative with a zero average value (Figure 3-5a). A 4th-order, low-pass, Butterworth filter modified the input as shown in Figure 3-5b. By selecting a higher order filter, the output drops more quickly to zero (Figure 3-5c). Low-pass filters are characterized by their 3-dB frequency and filter order (discussed in Section 9.5.1., *Low-pass Filter*). Electronic MTFs are typically plotted on a logarithmic scales whereas linear scales are used for electronic imaging systems.

74 SAMPLING, ALIASING, and DATA FIDELITY

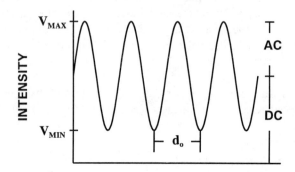

Figure 3-3. Definition of target modulation. The value d_o is the extent of one cycle.

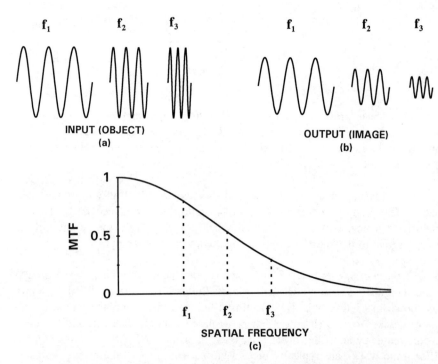

Figure 3-4. Modulation transfer function. (a) Input signal for three different spatial frequencies, (b) output for the three frequencies, and (c) the MTF is the ratio of output-to-input modulation.

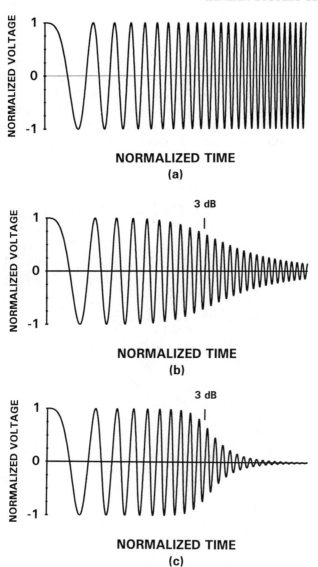

Figure 3-5. Input and output electronic signals. The input (a) is a sweep frequency signal where the sinusoidal frequency increases in time. It is identical to a frequency modulated (FM) signal. (b) The output of a 4th-order, low-pass, Butterworth filter. (c) The output of a 10th-order, low-pass, Butterworth filter.

76 SAMPLING, ALIASING, and DATA FIDELITY

Radiation intensity is always positive so that all scenes have an average value (Figure 3-6a). In Figure 3-6b, the intensity was modified by the optical MTF of a diffraction limited circular aperture. The MTF is equal to zero at and above the optical cutoff frequency. Electronic imaging systems can *detect* signals with frequencies above cutoff but cannot reproduce them as sinusoids. They just appear as a constant intensity (a value of 0.5 in Figure 3-6b). Generally, objects contain a distribution of frequencies. Detail is associated with high spatial frequencies. If these frequencies are above cutoff, the detail is lost. Here, a small target will appear as a blob.

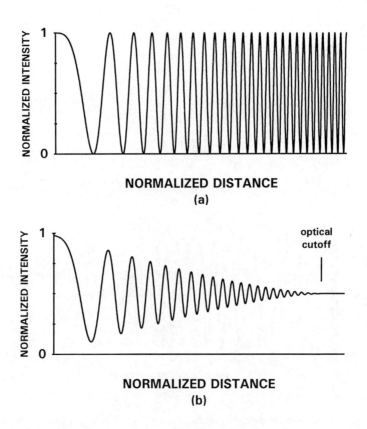

Figure 3-6. Input and output optical signals. (a) The input is a sweep frequency target where the sinusoidal frequency increases in distance. (b) For all optical systems, the output modulation is zero when the optical MTF is zero.

3.4. RESPONSE OF IDEALIZED CIRCUITS

Equation 3-21 (page 72) describes the dependence of the output signal on system MTF and PTF. The mathematics associated with practical components often becomes cumbersome. Idealized components are used to keep the mathematics relatively simple. These idealizations are physically unrealizable and therefore lead to physically unrealizable results. However, the results are quite useful for drawing general conclusions. But care must be exercised when interpreting the results.

All real electronic circuits introduce a certain amount of distortion because of bandwidth limitations and nonlinear phase characteristics. Idealized circuits will be used when discussing sampling theory. Real circuits are discussed in Chapter 9, *System Analysis*.

The ideal low-pass filter is always mentioned when discussing sampling theory. The phase shift is assumed linearly proportional to frequency. The ideal filter transfer function is

$$H(f) = Ae^{-2\pi t_d f} \quad \text{when } |f| < f_c$$
$$= 0 \quad \text{elsewhere} \quad (3\text{-}27)$$

where t_d is a constant of this circuit. This filter has two desired features. Band-limiting to f_c makes it an ideal anti-alias filter and reconstruction filter (discussed in Chapter 4, *Sampling*). For distortionless transmission though a circuit, the circuit must have linear phase and constant amplitude over the frequencies of interest.

The output of a circuit is $V_{OUT}(f) = V_{IN}(f)H(f)$. If the input is a single pulse centered on the origin, the output after passing through the ideal filter is

$$V_{OUT}(f) = V\tau\,\text{sinc}(\tau f)Ae^{-2\pi t_d f} \quad \text{when } |f| < f_c$$
$$= 0 \quad \text{elsewhere}. \quad (3\text{-}28)$$

78 SAMPLING, ALIASING, and DATA FIDELITY

The output pulse in time is

$$v_{OUT}(t) = \int_{-f_c}^{f_c} V_{OUT}(f) e^{2\pi t f} df. \qquad (3\text{-}29)$$

Using $e^{jx} = \cos(x) + j \sin(x)$ and noting that the integral of an odd function vanishes with symmetric limits provides

$$v_{OUT}(t) = AV\tau \int_{-f_c}^{f_c} sinc(\tau f) \cos[2\pi(t-t_d)f] df. \qquad (3\text{-}30)$$

While this cannot be evaluated in closed form, several features are notable.[2] The output will be delayed by t_d. The negative linear phase of the filter resulted in a time delay equal to the slope of the PTF. Further, the integral has a finite value for time before the pulse. That is, there is an output before the input. This violates causality and is a result of using the ideal filter characteristic. This also demonstrates that the ideal filter cannot be achieved with real circuit elements.

The Laplace transform is often used to describe electronic circuit response because it can accommodate initial (boundary) conditions. However, the Laplace transform is a ratio of polynomials that aides in locating poles and zeros. But for steady state conditions, the Fourier transform is appropriate. It describes the steady state response through the MTF and PTF.

3.5. SUPERPOSITION APPLIED TO OPTICAL SYSTEMS

If the system MTF is known, the image for any arbitrary object can be computed. First, the object is dissected into its constituent spatial frequencies (i.e., the Fourier transform of the object is obtained). Next, each of these frequencies is multiplied by the system MTF at that frequency. Then the inverse Fourier transform provides the image.

To illustrate the superposition principle and MTF approach, we will show how an ideal optical system modifies an image. An ideal optical system is, by definition, a linear-phase-shift system. The most popular test target consists of a series of bars: typically three or four bars although more may be used. For illustrative purposes, the periodic bars are assumed of infinite extent. A one-dimensional square wave, when expanded into a Fourier series about the origin, contains only odd harmonics. Using equations 2-22 (page 36) and 2-12 (page 32) for a square wave whose amplitude varies from zero to one,

$$o(x) = \frac{1}{2} + \frac{2}{\pi} \sum \frac{1}{n} \sin\left(\frac{2\pi n x}{d_o}\right) \quad n = 1, 3, 5, \cdots. \quad (3\text{-}31)$$

where d_o is the period of the square wave. The fundamental frequency u_o is $1/d_o$. Note that the peak-to-peak amplitude of the fundamental is $4/\pi$ times the square wave amplitude. In the frequency domain, the square wave provides discrete spatial frequencies of $u_n = 1/d_o$, $3/d_o$, $5/d_o$, $\cdots$ with amplitudes $2/\pi$, $2/3\pi$, $2/5\pi$, $\cdots$, respectively.

Let a circular optical system image the square wave. The MTF for a circular, clear aperture, diffraction-limited lens is

$$MTF_{OPTICS}(u) = \frac{2}{\pi}\left[\cos^{-1}\left(\frac{u}{u_{iC}}\right) - \frac{u}{u_{iC}}\sqrt{1 - \left(\frac{u}{u_{iC}}\right)^2}\right] \text{ when } u < u_{iC} \quad (3\text{-}32)$$

$$= 0 \quad elsewhere.$$

The optical cutoff in image space is $u_{iC} = D_o/(\lambda \, fl)$, where D_o is the aperture diameter, λ is the average wavelength, and fl is the focal length.

By superposition, the diffraction-limited optical MTF and square wave Fourier series amplitudes are multiplied together at each spatial frequency:

$$I(u_n) = MTF_{OPTICS}(u_n) O(u_n). \quad (3\text{-}33)$$

Taking the inverse Fourier transform provides the resultant image. Equivalently,

$$i(x) = \frac{1}{2} + MTF_{OPTICS}(u_o)\left[\frac{2}{\pi}\sin(2\pi u_o x)\right]$$
$$+ MTF_{OPTICS}(3u_o)\left[\frac{2}{3\pi}\sin(6\pi u_o x)\right] + \cdots \quad (3\text{-}34)$$

80 SAMPLING, ALIASING, and DATA FIDELITY

If u_o is greater than $u_{iC}/3$, only the fundamental of the square wave will be faithfully reproduced by the optical system. Here, the square wave will appear as a sine wave. Note that the optical MTF will reduce the image amplitude. As u_o decreases, the image will look more like a square wave (Figure 3-7). Because the optics does not pass any frequencies above u_{iC}, the resultant wave form is a truncation of the original series modified by the optical MTF. This results in some slight ringing. This ringing is a residual effect of the Gibbs phenomenon (see Section 2.1., *Fourier Series*, page 31). Note that with imaging systems, the intensity is always positive. As the input frequency approaches u_{iC} (Figure 3-7d), the modulation is barely present in the image. However, the average scene intensity approaches 0.5. That is, the scene detail can no longer be perceived at or above u_{iC}. *The scene did not disappear.*

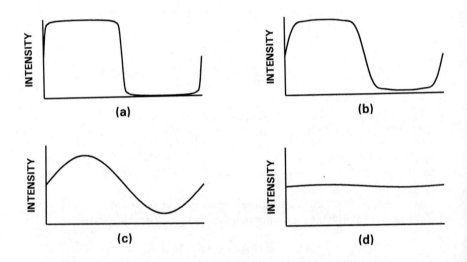

Figure 3-7. A square wave modified by a circular aperture as a function of x/d_o. As the square wave fundamental frequency increases, the edges become rounded. Eventually, the square wave will appear as a sinusoid when $u_o \geq u_{iC}/3$. (a) $u_o = u_{iC}/40$, (b) $u_o = u_{iC}/10$, (c) $u_o = u_{iC}/3$, and (d) $u_o = 9u_{iC}/10$.

3.6. REFERENCES

1. G. C. Holst, *Electro-Optical Imaging System Performance*, pp. 381-411, JCD Publishing, Winter Park, FL (1995).
2. M. Schwartz, *Information, Transmission, Modulation, and Noise*, second edition, pp. 56-64, McGraw-Hill, New York (1970).

4
SAMPLING

The sampling theorem, as introduced by Shannon was applied to information theory. He stated that if a time-varying function, v(t), contains no frequencies higher than f_{MAX} (Hz), it is completely determined by giving its ordinates at a series of points spaced $1/2f_{MAX}$ sec apart. The original function can be reconstructed by using an ideal low-pass filter.

Three conditions must be met to satisfy the sampling theorem. The signal must be band-limited, the digitizer must sample the signal at an adequate rate, and a low-pass reconstruction filter must be present. When any of these conditions are not present, the reconstructed analog signal may not be a replica of the original.

Most signals encountered with data acquisition systems and electronic imaging systems are not periodic. It is easily shown that any function can be decomposed into a series of sinusoidal frequencies. If these frequencies are below the Nyquist frequency (defined as one-half the sampling frequency), then a replica of the signal can be reconstructed. But aperiodic objects contain an infinite number of frequencies. Clearly, some of these frequencies will be above Nyquist frequency and will be aliased to lower frequencies. After reconstruction, the object will be distorted. To minimize aliasing, a low-pass filter (anti-alias filter) must be inserted before the sampler.

The zero-order reconstruction filter is the simplest and most often used filter. It is typically used for back-of-the-envelope diagrams. However, this filter passes frequencies above the Nyquist frequency and, therefore, is inadequate from a sampling theory view point. It creates a blocky signal. Other electronic reconstruction filters exist. This important topic is further discussed in Chapter 7, *Reconstruction*.

The symbols used in this book are summarized in the *Symbol List* (page xiii) which appears after the *Table of Contents*.

4.1. SAMPLED DATA SYSTEMS

Figure 4-1 illustrates an electronic c/d/c system. The analog signal is first sampled and then quantized. These two operations normally take place within an ADC. The sampled signal is then processed in a computer memory. It is converted to an analog signal (i.e., digital numbers are converted into a voltage) and then transformed into a continuous signal (reconstructed). Most digital-to-analog converters (DACs) also contain a sample-and-hold circuit to provide zero-order reconstruction. An additional reconstruction filter may be used to provide better signal fidelity. Sometimes, the ADC is called the modulator and the DAC is the demodulator.

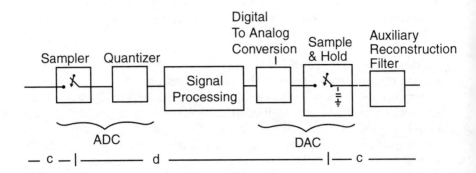

Figure 4-1. The electronic c/d/c system. ADCs and DACs provide two functions. They are (1) conversion between continuous and discrete data and (2) conversion between analog amplitude and quantized values.

Figure 4-2 illustrates an electronic imaging system. The scene is spatially sampled by the detectors. Each detector provides an analog output that may be modified by analog electronics. Then the signal is quantized by an ADC that may reside on the detector-chip assembly. Through clock signals, the now digital data are moved into a computer memory for image processing. This electronic image may be converted into an analog (video) signal which is fed into a monitor or may be directly sent to a digital monitor. The monitor and eye act together as reconstruction filters to create a perceived continuous image. The more complicated system, the c/d/c/d/c system (Figure 1-17, page 24) is just two c/d/c systems operating in series.

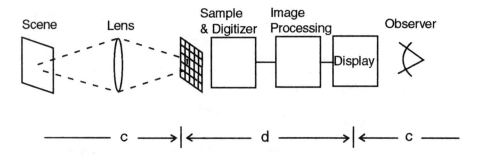

Figure 4-2. A simplified electronic imaging system shown as a c/d/c system. The digitizer may reside on the detector-chip assembly.

When using frequency domain analysis, one-dimensional analysis is used for the system illustrated in Figure 4-1 and two-dimensional analysis is used for the Figure 4-2 system. However, two-dimensional transforms are difficult to illustrate (they require three dimensions). Therefore, when describing an electronic imaging system response, a one-dimensional response is typically used (as often done in this text). Two notable differences exist as described below.

A staring array can significantly undersample the image whereas the sampler in Figure 4-1 may slightly undersample the signal. As such, a one-dimensional approach often understates the amount of aliasing present. The aliased signal must be correctly illustrated with the Nyquist frequency clearly labeled.

The popular zero-order reconstruction filter does not exist in most electronic imaging systems. For imagery, the display medium and the eye usually act as reconstruction filters. Since the reconstruction filters differ, the representation of data and output signals should also differ. However, there is a tendency to represent digital data using the zero-order filter. This may lead to misinterpretation of results.

4.2. SAMPLING THEORY

Figure 4-3 illustrates a band-limited, continuous, voltage that is sampled. If sampling occurs every T seconds, the sampling frequency is $f_S = 1/T$. The resultant signal is

$$v_{SAMPLE}(t) = v(t) \cdot s(t) , \qquad (4\text{-}1)$$

where s(t) is the sampling function equal to $\delta(t - nT)$. Since multiplication in one domain is represented as convolution in the other for LSI systems, the sampled frequency spectrum is

$$V_{SAMPLE}(f) = V(f) * S(f) , \qquad (4\text{-}2)$$

where V(f) is the amplitude spectrum of the band-limited analog signal and S(f) is the Fourier transform of the sampler. The transform S(f) is a series of impulses at $\pm nf_S$ and is called a comb function. When convolved with V(f), the resultant is a replication of V(f) about $\pm nf_S$. Equivalently, the sampling frequency interacts with the signal to create sum and difference frequencies. Any input frequency, f, will appear as $nf_S \pm f$ after sampling (Figure 4-3c):

$$V_{SAMPLE}(f) = \sum_{-\infty}^{\infty} V(nf_S \pm f) . \qquad (4\text{-}3)$$

Figure 4-4 illustrates a band-limited system with frequency components replicated by the sampling process. The base band ($-f_H$ to f_H) is replicated at nf_S. These replications are called side bands. To avoid distortion, the lowest possible sampling frequency is that value where the base band adjoins the first side band (Figure 4-4c). This is the graphical representation of the sampling theorem: a band-limited system must be sampled at twice the highest frequency ($f_S \geq 2f_H$) to avoid distortion in the reconstructed image.

After sampling, the signal is simply an array of digital numbers residing in a memory. The user assigns units during image reconstruction. For electronic circuitry, it is the clock timing that creates the units between sample points. For electronic imaging systems, it is the relationship between the data array size and the camera field-of-view.

SAMPLING THEORY 85

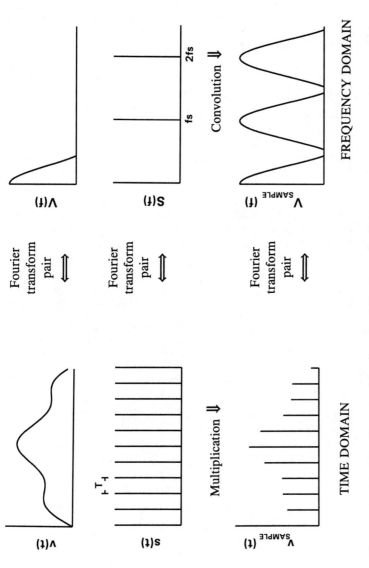

Figure 4-3. Sampling creates replications in the frequency domain. Only the positive frequencies are illustrated. The transform pairs are $v(t) \Leftrightarrow V(f)$, $s(t) \Leftrightarrow S(f)$, and $v_{SAMPLE}(t) \Leftrightarrow V_{SAMPLE}(f)$.

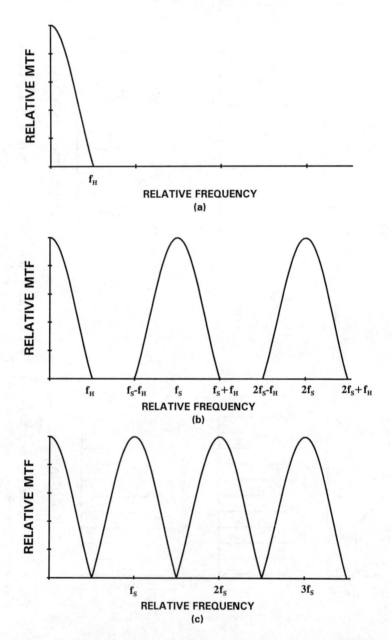

Figure 4-4. Sampling replicated frequencies at $nf_S \pm f$. (a) Original band-limited signal. (b) Amplitude spectrum after sampling. (c) When $f_S = 2f_H$, the bands just adjoin. Only the positive frequencies are illustrated.

4.3. RECONSTRUCTION FILTER

The reconstruction filter converts the digital data (which we cannot see) into an analog signal (which we can see). It limits the frequency content of the reconstructed analog signal to

$$V_{RECON}(f) = V_{SAMPLE}(f)H_{RECON}(f) = H_{RECON}(f)\sum_{-\infty}^{\infty} V(nf_s \pm f) . \quad (4\text{-}4)$$

Since the response is symmetrical about zero, using only positive frequencies makes interpretation easier:

$$V_{RECON}(f) = V(f)H_{RECON}(f) + H_{RECON}(f)\sum_{1}^{\infty} V(nf_s \pm f) . \quad (4\text{-}5)$$

The first term is just the original signal modified by the reconstruction filter response. The second term was created by the sampling process. If any portion of these frequencies remains in the analog signal, the reconstructed analog signal will be a distorted version of the original signal.

The ideal filter will have unity response up to f_N and then zero after that (Figure 4-5). If the original signal was oversampled ($f_N > f_H$) and if the reconstruction filter limits frequencies to f_N, then the reconstructed image can be identical to the original image. That is, the second term of Equation 4-5 is zero.

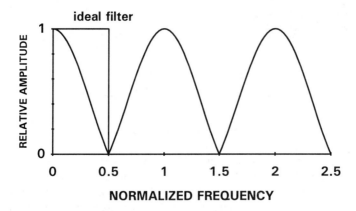

Figure 4-5. Ideal reconstruction filter. $f_S = 1$ and $f_N = 0.5$.

88 SAMPLING, ALIASING, and DATA FIDELITY

The ideal filter is, of course, unrealizable. A non-ideal filter can create an amplitude spectrum that contains frequencies not in the original signal. That is, $H_{RECON}(f)$ passes part of the frequencies represented by the second term of Equation 4-5. The zero-order filter is the most common for back-of-the-envelope diagrams and is found in most DACs. The digital data are converted to an analog signal and held constant until the next digital signal is available (sample-and-hold function). The zero-order filter response is

$$H_{RECON}(f) = \left| sinc\left(\frac{f}{f_s}\right) \right| . \qquad (4\text{-}6)$$

Figure 4-6 illustrates the zero-order reconstruction filter. As illustrated in Figure 4-7, the resultant analog signal is blocky. This is often called the pixellation effect.

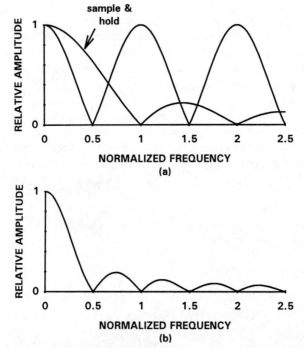

Figure 4-6. (a) Replicated amplitude spectra and the zero-order reconstruction filter. (b) Resultant amplitude spectrum after the zero-order filter. The remnants above $f = 0.5$ (the Nyquist frequency) create the blocky image. $f_s = 1$. Only the positive frequencies are illustrated.

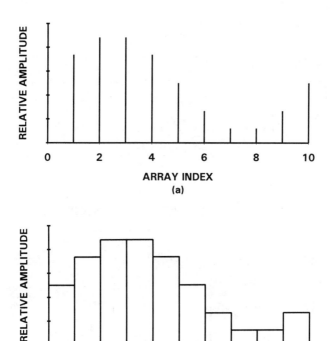

Figure 4-7. The zero-order reconstruction filter creates a blocky signal. (a) Digital signal representation and (b) analog signal after a zero-order filter.

4.4. ALIASING

As the sampling frequency decreases, the first side band starts to overlap the base band (Figure 4-8). The overlaid region creates distortion in the reconstructed signal. Figure 4-9 illustrates the replication of f_o at $f_S - f_o$. When using an ideal reconstruction filter, it is impossible to tell whether the reconstructed frequency resulted from an input frequency of f_o or $nf_S \pm f_o$ (Figure 4-10). This is aliasing. Once aliasing has occurred, it cannot be removed and the original signal can never be faithfully recovered.

90 SAMPLING, ALIASING, and DATA FIDELITY

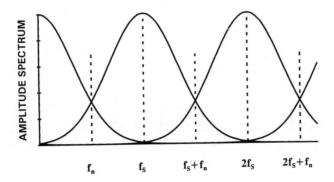

Figure 4-8. Frequency overlap occurs with undersampled systems.

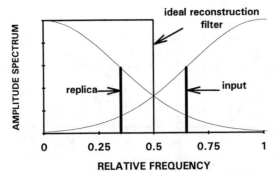

Figure 4-9. An input above Nyquist frequency will be replicated below Nyquist frequency. The ideal reconstruction filter will pass this replicated frequency. $f_S = 1$ and $f_N = 0.5$.

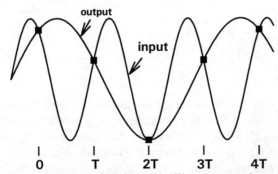

Figure 4-10. An undersampled sinusoid will appear as a lower frequency after an ideal reconstruction filter. The heavy dots are the samples taken at time nT. The sampling frequency is $f_S = 1/T$. The input frequency is $1/1.43T$ and the output appears as $1/3.33T$. Nyquist frequency is $1/2T$.

4.5. ANTI-ALIAS FILTER

In some situations, the sampling frequency is fixed and then the input frequency must be limited by an anti-alias filter to avoid aliasing. Without this filter, some signals will be undersampled. Figure 4-11 illustrates the cutoff features of an ideal filter. The ideal filter passes all frequencies below f_N without modification and attenuates the amplitude of all frequencies above f_N. Most electronic test equipment and frame grabbers do not have an anti-alias filter.

A very sharp cutoff filter may introduce ringing in the output (see Figure 2.6, page 37). Whether aliasing is considered undesirable depends on the application. Anti-alias filter design trades off acceptable ringing with attenuation in the pass band (discussed in Section 9.5.1., *Low-pass Filter*).

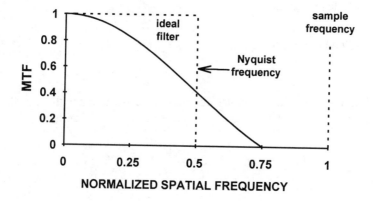

Figure 4-11. Ideal anti-alias filter. Although the filter characteristics are identical to the reconstruction filter, its purpose is to minimize aliasing. The ideal anti-alias filter passes all the signals below f_N and no signal above f_N.

92 SAMPLING, ALIASING, and DATA FIDELITY

4.6. TWO-DIMENSIONAL SAMPLING

The extension of one-dimensional to two-dimensional sampling is straightforward. The most popular detector array configuration is rectangular and it creates a rectangular sampling lattice. Usually the horizontal and vertical directions are considered separable so that sampling theory can be applied to each direction separately:

$$O(u,v) = O(u)O(v) . \qquad (4\text{-}7)$$

The one-dimensional figures shown though this text (e.g., Figure 4-5) can also represent the frequency response along the u or v axes (Figure 4-12). This provides the relationship between one-dimensional and two-dimensional sampling theory.

The sampled signal is

$$I(u,v) = O(u,v) ** S(u,v) = \sum_{n=-\infty}^{\infty} \sum_{m=-\infty}^{\infty} O(nu_S \pm u, mv_S \pm v) . \qquad (4\text{-}8)$$

Figure 4-13 illustrates the two-dimensional frequency spectrum with replications at the sampling frequencies, u_S and v_S. The reconstructed image is

$$I_{RECON}(u,v) = F_{RECON}(u,v) \sum_{n=-\infty}^{\infty} \sum_{m=-\infty}^{\infty} O(nu_S \pm u, mv_S \pm v) . \qquad (4\text{-}9)$$

Using only positive frequencies

$$I_{RECON}(u,v) = F_{RECON}(u,v) O(u,v)$$

$$+ F_{RECON}(u,v) \sum_{n=1}^{\infty} \sum_{m=1}^{\infty} O(nu_S \pm u, mv_S \pm v) . \qquad (4\text{-}10)$$

As shown in Figure 4-14, the ideal two-dimensional reconstruction filter should just enclose the area specified by the two Nyquist frequencies:

$$F_{RECON}(u,v) = 1 \quad \text{when } |u| < u_N \text{ and } |v| < v_N$$
$$= 0 \quad \text{elsewhere} . \qquad (4\text{-}11)$$

If $O(u,v)$ is band-limited to u_N and v_N then the ideal reconstruction filter will return $O(u,v)$ and the second term of Equation 4-10 becomes zero.

As with the one-dimensional case, the two-dimensional ideal filter is physically unrealizable. For electronic imaging systems, two reconstruction filters are typically used: the display medium and the eye. While a flat panel display may act as a zero-order reconstruction filter, neither a CRT-based monitor nor the eye does. These filters are further discussed in Chapter 7, *Reconstruction*.

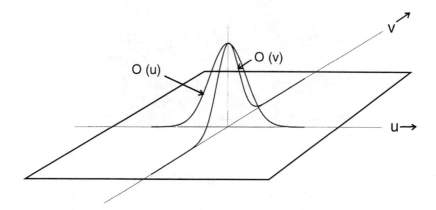

Figure 4-12. Separable frequency components. $O(u)$ is the amplitude spectrum along the u-axis and $O(v)$ is the amplitude spectrum along the v-axis. The two-dimensional spectrum is $O(u)O(v)$.

94 SAMPLING, ALIASING, and DATA FIDELITY

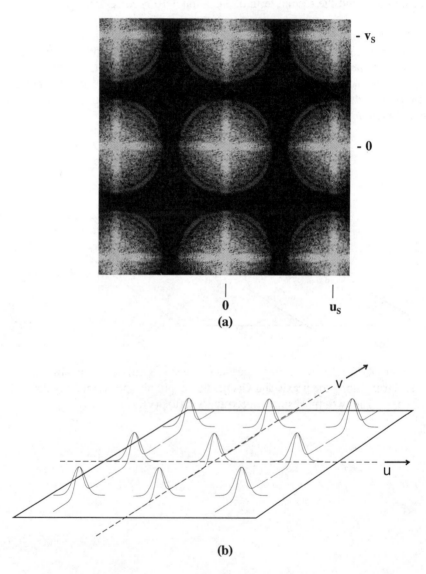

Figure 4-13. With two-dimensional sampling, the frequency spectrum is replicated about $\pm nu_s \pm mv_s$. (a) Two-dimensional representation where gray levels indicate amplitude. (b) Contour plot.

SAMPLING THEORY 95

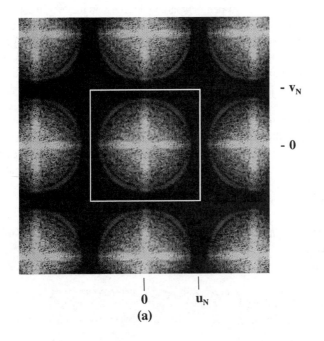

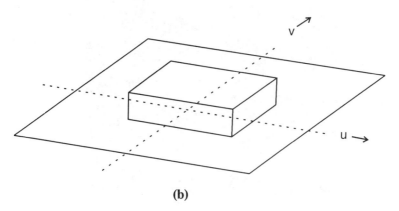

Figure 4-14. Ideal two-dimensional reconstruction filter. (a) Two-dimensional representation where gray levels indicate amplitude. The box defines the Nyquist frequency in two dimensions. (b) Contour plot of the ideal filter.

96 SAMPLING, ALIASING, and DATA FIDELITY

The effective detector center-to-center spacing determines the sampling frequency. As the center-to-center spacing increases, the sampling frequencies decrease (Figure 4-15). If the spacing is too large, two-dimensional aliasing occurs (Figure 4-16). Undersampling creates moiré patterns in imagery.

Since aliasing occurs at the detector, it can only be avoided by optically limiting the signal. The ideal anti-aliasing filter will attenuate all frequencies above the Nyquist frequency. Optical band-limiting can be achieved by using small diameter optics, defocusing, blurring the image, or by inserting a birefringent crystal between the lens and array. The birefringent crystal changes the effective sampling rate and is found in almost all single chip color cameras[1] (discussed in Section 5.3.8., *Optical Prefiltering*). Unfortunately, these approaches also degrade the MTF and reduce image sharpness. Most electronic imaging systems do not have an anti-alias filter.

An electronic anti-alias filter cannot remove the aliasing that has taken place at the detector. It can only minimize further aliasing that might occur in any downstream analog-to-digital converter.

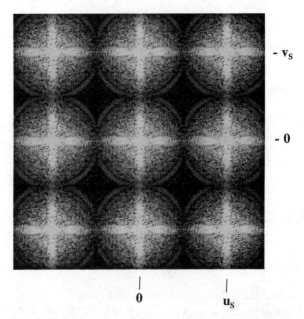

Figure 4-15. Increasing the center-to-center spacing decreases the sampling frequency. This brings the spectra closer together.

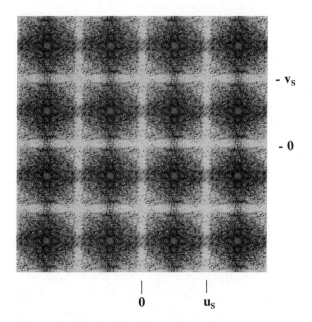

Figure 4-16. When u_N and v_N are smaller than the highest frequencies present, spectra overlap and aliasing occurs.

4.7. REFERENCES

1. J. E. Greivenkamp, "Color Dependent Optical Prefilter for the Suppression of Aliasing Artifacts," *Applied Optics*, Vol. 29(5), pp. 676-684 (1990).

5

SAMPLING DEVICES

For electronic circuits, the analog-to-digital converter is the sampling device. It operates on a continuous analog signal (assumed to be a voltage) to create a one-dimensional representation of the voltage-time history.

For electronic imaging systems, the detector locations create a two-dimensional sampling lattice. The detector output is an instantaneous representation of the scene intensity. If the effective detector center-to-center spacings are different in the horizontal and vertical directions, sampling effects will be different in the two directions. Electronic imaging systems are complex and several different samplers may exist. If the camera output is in an analog format, the output may again be digitized by a frame grabber. See, for example, the c/d/c/d/c system illustrated in Figure 1-17 (page 24).

In previous chapters, it was implied that the sampler was ideal. Real samplers have a finite aperture over which the signal is integrated. This integration is represented by the sampler's MTF.

The symbols used in this book are summarized in the *Symbol List* (page xiii) which appears after the *Table of Contents*.

5.1. ANALOG MULTIPLEXER

An ideal sampler was used (see Figure 4-3, page 85) to develop sampling theory. It provided an instantaneous representation of the signal. That is, the sampled pulse width was essentially zero. Analog samplers have a finite aperture over which they pass the signal. These are analog multiplexers and are often used to multiplex several analog signals prior to digitization. They may also be considered as the initial component of an ADC (see Figure 1-2, page 4).

SAMPLING DEVICES 99

Referring to Figure 5-1, the signal is band-limited to $f_{MAX} < f_N$ where $f_N = 1/2T$. The sampled signal is

$$V_{SAMPLE}(t) = v(t)s(t) . \qquad (5\text{-}1)$$

Because multiplication in one domain is represented as convolution in the other, the sampled frequency spectrum is

$$V_{SAMPLE}(f) = V(f) * S(f) . \qquad (5\text{-}2)$$

The Fourier transform of the sampler with an aperture width of τ is

$$S(f) = sinc(\tau f) \, \delta(f \pm nf_S) = sinc(\tau f) comb\left(\frac{f}{f_S}\right) . \qquad (5\text{-}3)$$

The transform is nonzero when $f = \pm nf_S$ and is normalized to unity at $f = 0$. When convolved with V(f), the result is a replication of V(f) about $\pm nf_S$ with an amplitude that decreases with sinc(τf). The magnitude, $|sinc(\tau f)|$, may be considered the analog multiplexer MTF. As the window width decreases, the first zero of $|sinc(\tau f)|$ increases in frequency and thereby incorporates more frequency terms.

The idealized situation occurs when $\tau \to 0$. The multiplexer MTF approaches unity, $S(f) \to \delta(f \pm nf_S)$, and all frequency components are present from $-\infty$ to ∞. While the idealized case is physically unrealizable, it simplifies the mathematics. This MTF only affects the frequency replications about $\pm nf_S$. Because the reconstruction filter only passes the base band and sometimes part of the first few replicated bands, letting $\tau = 0$ will not significantly affect the reconstructed signal appearance.

In the limit where $\tau = T$, the signal becomes continuous. The zeros of $|sinc(\tau f)|$ exactly occur at nf_S and the only component of S(f) remaining exists at the origin and $S(f) = \delta(f)$. Convolving S(f) with V(f) just returns V(f).

100 SAMPLING, ALIASING, and DATA FIDELITY

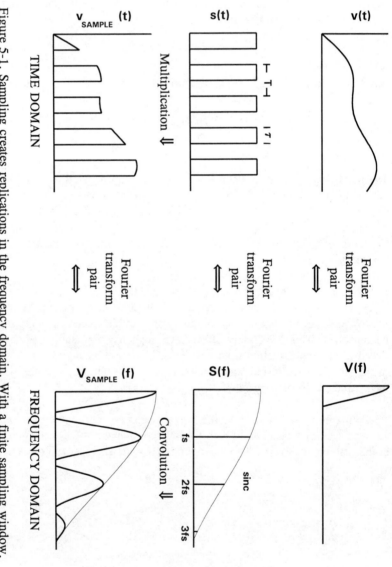

Figure 5-1. Sampling creates replications in the frequency domain. With a finite sampling window, the frequency spectrum decreases as a sinc function. Only positive frequencies are shown. The transform pairs are $v(t) \Leftrightarrow V(f)$, $s(t) \Leftrightarrow S(f)$, and $v_{SAMPLE}(t) \Leftrightarrow V_{SAMPLE}(f)$.

5.2. ANALOG-TO-DIGITAL CONVERTERS

ADCs are used in all test equipment and frame grabbers. They convert an analog time-varying signal into a digital signal. Referring to Figures 1-15 though 1-17 (pages 22-24) they exist at every c/d interface including detectors. Detectors produce an analog output that is digitized by an ADC.

An ADC integrates the signal over the aperture width of the analog sampler. This integration produces a new function, $v_{AVE}(t)$, which is the running average of v(t):

$$v_{AVE}(t) = \int_{-\infty}^{\infty} v(t') \, rect\left(\frac{t-t'}{\tau}\right) dt' . \quad (5\text{-}4)$$

Because it is the same form as the convolution integral (Equation 3-7, page 67), the normalized Fourier transform is

$$V_{AVE}(f) = sinc(\tau f) V(f) , \quad (5\text{-}5)$$

where $|sinc(\tau f)|$ may be considered the MTF of the ADC. As illustrated in Figure 5-2, if $\tau \ll T$, then $MTF_{ADC}(f)$ does not significantly affect V(f). ADCs that operate in this fashion are called flash converters.

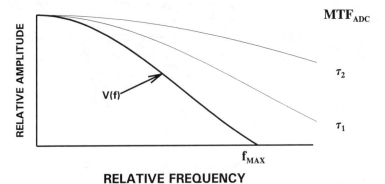

Figure 5-2. Frequency response of a typical ADC. The signal is band-limited to f_{MAX} where $f_{MAX} < 1/2T$. As τ becomes smaller, $MTF_{ADC}(f)$ increases ($\tau_2 < \tau_1$). The response is $V(f)MTF_{ADC}(f)$.

102 SAMPLING, ALIASING, and DATA FIDELITY

Here, s(t) provides a sample every T sec:

$$s(t) = comb\left(\frac{t}{T}\right). \tag{5-6}$$

The function $V_{AVE}(f)$ is convolved with $comb(f/f_S)$ to obtain $V_{SAMPLE}(f)$. In the idealized case, $\tau = 0$ and $MTF_{ADC}(f)$ is unity for all frequencies from $-\infty$ to ∞. While physically unrealizable, it simplifies the mathematics. Because the reconstruction filter only passes the base band and sometimes part of the first few replicated bands, letting $\tau = 0$ will not significantly affect the reconstructed signal appearance.

5.3. DETECTORS

The detector is the heart of every electronic imaging system because it converts scene radiation into a measurable electrical signal. As a linear component, it continuously converts photons into electrons. Figure 1-3 (page 4) illustrated how a detector array samples a scene.

5.3.1. DETECTOR MTF

The detector integrates the signal over its active area. Figure 5-3 illustrates a one-dimensional sinusoid viewed by a detector. The sinusoid moves continuously and the detector analog output is continuously measured. Equivalently, a movable mirror could allow the detector-angular-subtense to scan along the sinusoid. For a detector whose photosensitive area is $d_H \times d_V$ mm^2, the analog output is

$$v_{AVE}(x) = R d_V \int_{-\infty}^{\infty} i(x') rect\left(\frac{x - x'}{d_H}\right) dx', \tag{5-7}$$

where R is the detector responsivity that relates output voltage to input intensity where the intensity is averaged over the active area and $i(x')$ is the image intensity variation falling onto the detector.

SAMPLING DEVICES 103

The integral is in the same form as the convolution integral, and the Fourier transform is

$$I_{AVE}(u_i) = sinc(d_H u_i) I(u_i) , \qquad (5\text{-}8)$$

where u_i is the image spatial frequency in cycles/mm. For this response to exist, a lens must image the sinusoidal pattern onto the detector. The scene sinusoidal modulation is, of course, modified by the optical MTF. This was described in Section 3.5., *Superposition Applied to Optical Systems* (page 78). In one dimension,

$$I(u_i) = O(u_i) MTF_{OPTICS}(u_i) . \qquad (5\text{-}9)$$

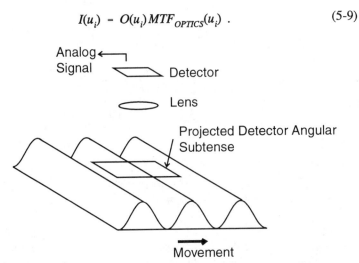

Figure 5-3. Detector scanning a one-dimensional sinusoid. It senses the radiation within the projected footprint. A detector requires a lens to resolve imagery.

Figure 5-4 illustrates the output for two different input frequencies. As $d_H u_i$ increases, the detector analog output decreases. When $d_H u_i = 1$, the output is zero. For $1 < d_H u_i < 2$, the output modulation is reversed from the input modulation. Here, dark bars appear as white bars. Because the MTF is always positive, the MTF is

$$MTF_{DETECTOR}(u_i) = \left| sinc(d_H u_i) \right| . \qquad (5\text{-}10)$$

The MTF and PTF are shown in Figure 5-5. Reversal of white and black bars appears as a π radian phase shift.

104 SAMPLING, ALIASING, and DATA FIDELITY

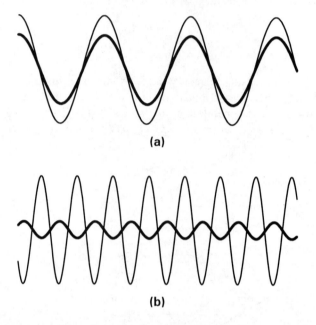

Figure 5-4. The light line is the image input and the heavy line is the detector analog output. (a) $d_H u_i = 0.5$ and (b) $d_H u_i = 1.2$.

The MTF is zero when $d_H u_i = k$. The first zero (k=1) is considered the detector cutoff ($u_{iD} = 1/d_H$). It is customary to plot the MTF only up to the first zero (Figure 5-6). Unfortunately, this representation may lead the analyst to believe no response occurs beyond u_{iD}. The terminology, *cutoff*, supports this interpretation. As a result of this common (and misleading) representation, some analysts state that when u_{iN} is equal to u_{iD}, the sampling theorem has been satisfied and that the image is appropriately sampled. Nevertheless, in the next few sections, the figures will show $MTF_{DETECTOR}$ plotted only up to u_{iD} for brevity.

When extended to two dimensions, the (separable) MTF is

$$MTF_{DETECTOR}(u_i,v_i) = |sinc(d_H u_i)| |sinc(d_V v_i)|, \quad (5\text{-}11)$$

where v_i is the vertical spatial frequency in image space. Although most authors treat the two dimensions as separable (also done in this text), recent work[1] indicates that the two-dimensional MTF is angularly dependent.

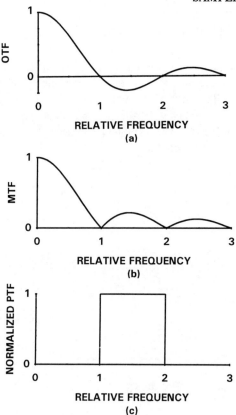

Figure 5-5. Detector response as a function of u_i/u_{iD}. (a) OTF, (b) MTF, and (c) PTF. The PTF amplitude is normalized to π. Only positive frequencies are shown. The response is symmetric about $u_i = 0$.

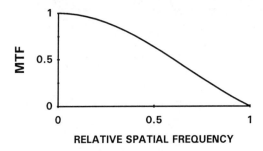

Figure 5-6. Typical detector MTF representation as a function of u_i/u_{iD}. The MTF is often plotted only up to the first zero. The response is symmetric about $u_i = 0$. This representation suggests that the detector has no response beyond u_{iD}.

5.3.2. DETECTOR ARRAY OUTPUT

Normally, the detectors are placed in an array, and the scene is static. Figure 5-7 illustrates the spatial integration afforded by a linear array of detectors. In Figure 5-7a, the detector width is one-half of the center-to-center spacing (detector pitch). That is, only 50% of the input sinusoid is detected. The detector integrates the signal over the range indicated and the heavy lines are proportional to the detector outputs.

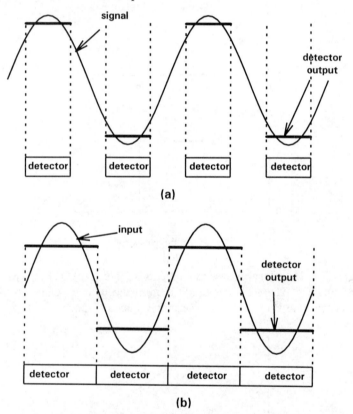

Figure 5-7. A detector spatially integrates the signal. The signal is a spatial sine wave. (a) Detector width is one-half of the center-to-center spacing and (b) detector width is equal to the center-to-center spacing. (b) typifies frame transfer arrays[2] and (a) typifies the horizontal arrangement of an interline transfer[3] CCD array. The heavy lines are the detector output voltages and represent the average image intensity across the detector. As the phase changes, the output also changes.

The output MTF is $(V_{MAX} - V_{MIN})/(V_{MAX} + V_{MIN})$, where V_{MAX} and V_{MIN} are the maximum and minimum voltages, respectively. As the detector size increases, it integrates over a larger region and the MTF decreases. As the phase changes between the detector location and the peak of the sinusoid, the output voltages change (discussed in Section 8.5., *Bar Target Appearance*). This creates a variable MTF that is a sampling artifact (discussed in Section 9.4., *Sample-scene MTF*).

The detector provides the interface between space and time. Spatial information is converted into a time-varying voltage. In a sampled data system, the detector analog output is digitized by a flash ADC. As discussed in Section 5.2, allowing the ADC window width approach zero simplifies the mathematics and data representation. For the remainder of this text, a zero width aperture is considered. Digitized data (collected temporally) is placed into a data array (Figure 5-8).

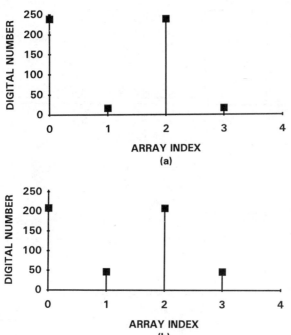

Figure 5-8. Graphical representation of sampled data from an 8-bit system. Sampled output for detector output shown in (a) Figure 5-7a and (b) Figure 5-7b. The digitized data reside in a computer memory and are identified by indices.

108 SAMPLING, ALIASING, and DATA FIDELITY

Because the data rate is known, the temporal data can be directly related to spatial coordinates. Figure 5-9 illustrates the back-of-the-envelope representation after the data have been clocked out of the data array. Note that in Figure 5-9a the output appears continuous even though the detectors spatially integrated a small part of the signal. When the detectors are not contiguous, small objects, whose image size is smaller than the dead space between detectors, will "disappear" when the image falls in the dead space. Small objects will "twinkle" as the image moves from one detector, onto the dead space, and then onto the next detector. The displayed image is always represented as a continuous signal. This blocky representation occurs with a sample-and-hold reconstruction filter. An ideal low-pass reconstruction filter will remove the blockiness and produce a sinusoid (given by Equation 5-10, page 103). It cannot prevent the twinkling of small objects.

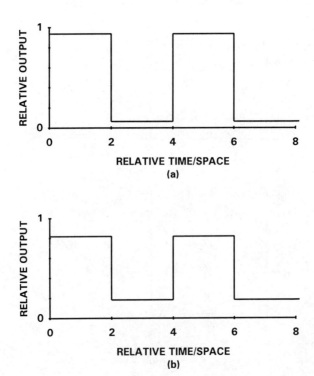

Figure 5-9. Sample-and-hold representation for the detector output shown in (a) Figure 5-8a and (b) Figure 5-8b. An ideal reconstruction filter will produce a sinusoid.

SAMPLING DEVICES 109

5.3.3. NYQUIST FREQUENCY

The discrete location of detectors in an array act as a sampling lattice whose sampling rate is $u_{iS} = 1/d_{CCH}$ and $v_{iS} = 1/d_{CCV}$. The values d_{CCH} and d_{CCV} are the effective detector center-to-center spacings (detector pitch). The Nyquist frequencies are $u_{iN} = 1/(2d_{CCH})$ and $v_{iN} = 1/(2d_{CCV})$. If $I(u_i,v_i)$ is limited to the Nyquist frequencies and an ideal low-pass reconstruction filter is used, the displayed image will be nearly the same as the original image. Note that the displayed image will be modified by the detector MTF. Because the optical MTF modifies the scene, the displayed image frequencies will differ from the scene frequencies by $MTF_{OPTICS}MTF_{DETECTOR}$.

Because of detector location symmetry, frame transfer[2] monochrome CCDs and many infrared imaging staring arrays tend to have equal sampling rates in both the horizontal and vertical directions. Figure 5-10 illustrates the MTFs associated with a staring array where $d_{CCH} = d_{CCV} = d_H = d_V$.

Figure 5-11 typifies a CCD interline transfer array.[3] The detectors are contiguous in the vertical direction but are separated in the horizontal direction. The potential aliasing can be different in the two directions.

Most single-chip color filter arrays (CFAs) have an unequal number of red, green, and blue detectors.[4] A typical array designed for NTSC operation will have 768 detectors in the horizontal direction with 384 detectors sensitive to green, 192 sensitive to red, and 192 sensitive to blue regions of the spectrum. Suppose the arrangement is G-R-G-B-G-R-G-B (Figure 5-12).

The horizontal spacing of the "blue" and "red" detectors are twice the "green" detector spacing. This produces a "blue" and "red" array Nyquist frequency that is one-half the "green" array Nyquist frequency (Figure 5-13). Other filter array layouts will create different array Nyquist frequencies. The "color" Nyquist frequencies can be different in the horizontal and vertical directions. These unequal array Nyquist frequencies create color aliasing in single-chip cameras that is wavelength specific.

The "color pixel" must be fully illuminated to provide the full color gamut. Suppose the image is only one detector wide. As the image moves across the color pixel, the reconstructed image will appear green, then red, green again, and finally blue. As the image continues to move, the color pattern will repeat. That is, black-and-white scenes can appear as green, red, or blue imagery.[5] A birefringent crystal, inserted between the lens and the array, effectively blurs the image so that a full color pixel is illuminated. This changes the effective detector size and minimizes color aliasing (See in Section 5.3.8., *Optical Prefiltering*).

110 *SAMPLING, ALIASING, and DATA FIDELITY*

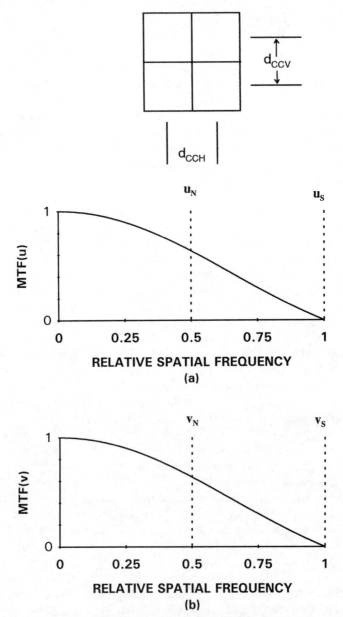

Figure 5-10. Staring array with contiguous square detectors. (a) MTF(u) and (b) MTF(v). The frequency axes are normalized to the detector cutoffs. This typifies CCD frame transfer devices.

SAMPLING DEVICES 111

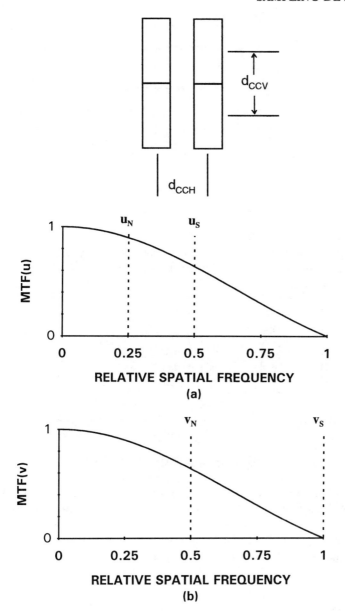

Figure 5-11. Staring array contiguous detectors in the vertical direction but separated in the horizontal direction ($d_{CCH} = 2d_H$). (a) MTF(u) and (b) MTF(v). This typifies interline transfer CCD arrays. The frequency axes are normalized to the detector cutoff frequencies.

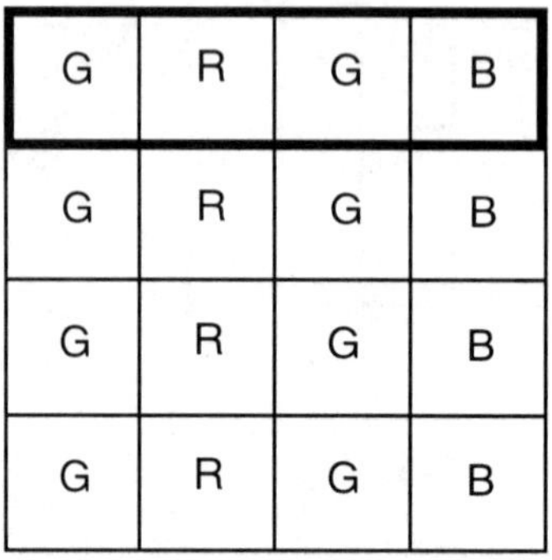

G	R	G	B
G	R	G	B
G	R	G	B
G	R	G	B

Figure 5-12. Typical arrangement of a color filter array. The heavy line outlines the "color pixel." The horizontal spacing between the "red" and "blue" detectors are twice the "green" detector spacing.

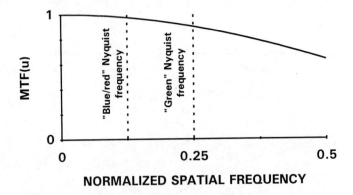

Figure 5-13. Detector horizontal MTF normalized to u_i/u_{iD} for the CFA illustrated in Figure 5-12. Unequally spaced red, green, and blue sensitive detectors create different array Nyquist frequencies. This creates different amounts of aliasing. The vertical MTFs are identical to the vertical MTFs illustrated in Figure 5-11.

5.3.4. MICROSCAN

For staring arrays, the sampling frequency is created by the detector center-to-center spacing. The effective sampling frequency can be increased[6-10] with a microscan or dither technique (also called microdither). Here, the image (through mechanical or optical means) moves a fraction of d_{CCH} and d_{CCV}. At each location, the detector stares at the image (Figure 5-14). Microscan can be either in one direction or in two directions. Microscan increases the sampling frequency and thereby permits faithful reproduction of higher spatial frequencies (Figure 5-15). The data are digitally interwoven in the computer memory. A similar effect occurs when the image is moving (discussed in Section 8.9., *Dynamic Sampling*).

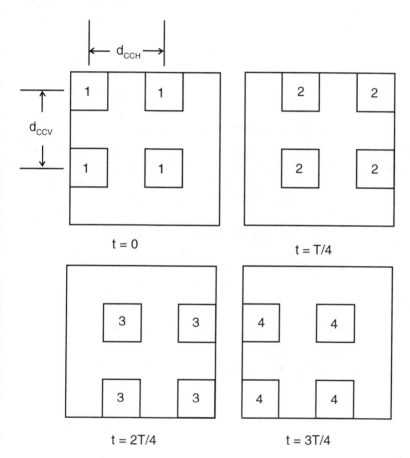

Figure 5-14. 2 × 2 microscan concept. The detector samples the image at four different locations. Four individual images are collected, each offset by $d_{CCH}/2$ and $d_{CCV}/2$. The reassembled image is a composite of the four samples. If the entire frame is produced every T seconds, then each subframe must be collected every T/4 sec.

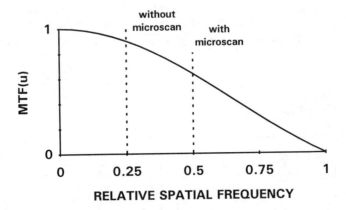

Figure 5-15. Microscan increases the array Nyquist frequency by decreasing the effective detector locations. The detectors are separated by $2d_H$ and $2d_V$. Image samples are separated by d_H and d_V. The detector width is $d_H = d$.

5.3.5. INFINITE STARING ARRAYS

When aliasing is included, the one-dimensional amplitude spectrum becomes

$$I_{INFINITE}(u_i) = \left(I(u_i) \cdot |sinc(d_H u_i)|\right) * \delta(nu_{is} \pm u_i)$$

$$= \left(I(u_i) \cdot |sinc(d_H u_i)|\right) * comb\left(\frac{u_i}{u_{is}}\right).$$

(5-12)

Figure 5-16 illustrates the replicated spectra when $d_{CCH} = d_H$. This overlap was portrayed in two dimensions in Figure 4-16 (page 97). In Figure 5-16b the spectrum up to the Nyquist frequency is shown. In Figure 5-17, $d_{CCH} = 2d_H$. The lower Nyquist frequency aliases more signal. The amount of aliased signal depends on the scene content and optical prefiltering. Because aliased signal is scene dependent, it cannot be predicted in advance. These figures show worst case: the signal exists for all frequencies.

The aliased signal can also be portrayed by directly integrating over the sinusoidal pattern. In Figure 5-18, the signal is sampled by four detectors. The output is digitized as shown in Figure 5-8 (page 107) and then passed though an ideal reconstruction filter. This returns all signals whose frequency components are in the base band. When the number of samples per period is less than two, the reconstructed signal appears as a lower frequency (aliased component).

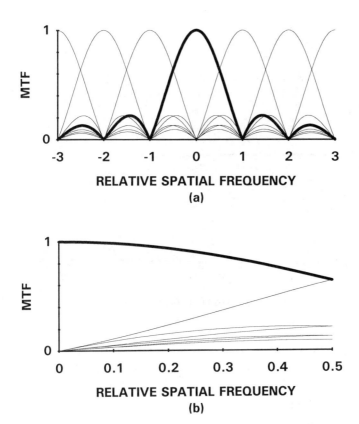

Figure 5-16. Replicated spectrum when $d_{CCH} = d_H$ as a function of u_i/u_{iD}. The heavy line is the detector MTF and the light lines are the replicated spectra. (a) Full spectrum when $|n| = 0, 1, 2$, and 3. (b) Spectrum up to u_{iN}.

116 *SAMPLING, ALIASING, and DATA FIDELITY*

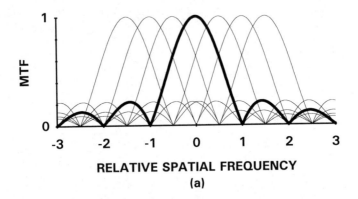

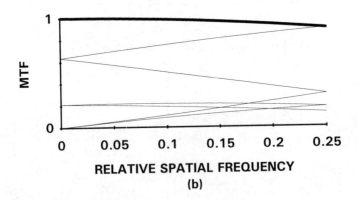

Figure 5-17. Replicated spectrum when $d_{CCH} = 2d_H$ as a function of u_i/u_{iD}. The heavy line is the detector MTF and the light lines are the replicated spectra. (a) Full spectrum when $|n| = 0, 1, 2,$ and 3. (b) Spectrum up to u_{iN}.

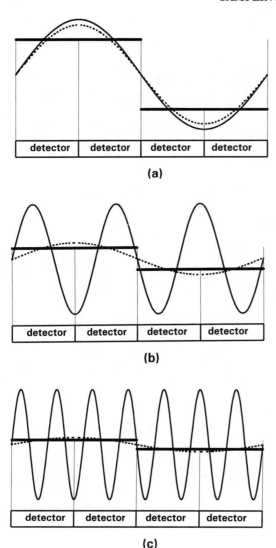

Figure 5-18. The heavy horizontal line is the detector output voltage and is proportional to the integrated intensity. The dashed line is the signal reconstructed by an ideal low-pass filter. (a) Four samples per period, (b) 1.33 samples per period, and (c) 0.57 samples per period. Equivalently, u_o/u_{is} is 0.25, 0.75, and 1.75, respectively. In (b) and (c), the signals are aliased to 0.25 and this is the same input frequency as (a). The amplitudes are governed by $MTF_{DETECTOR}$.

118 SAMPLING, ALIASING, and DATA FIDELITY

5.3.6. INFINITE SCANNING ARRAYS

With scanning systems, the detector output in the scan direction can be electronically digitized at any rate. In the cross scan direction, the detector locations define the sampling rate. Therefore, in a scanning system, the sampling rate may be significantly different in the horizontal and vertical directions. Scanning systems can have higher Nyquist frequencies than staring systems.

Figure 5-19 illustrates three different detector configurations. Figure 5-19a illustrates a single linear array whereas Figure 5-19b portrays two linear arrays offset as shown. Similarly, Figure 5-19c represents three linear arrays offset by $d_{CCV}/3$. By offsetting the detectors, any vertical Nyquist frequency can be created (Figure 5-20). Because of the misleading representation of the detector MTF (Figure 5-6, page 105), some analysts say that Figure 5-19b satisfies the Nyquist criterion. As a result, it is the most prevalent design.

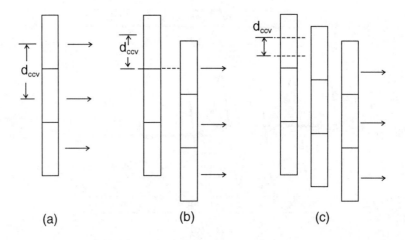

Figure 5-19. Three different scanning arrays. The scan direction is typically in the horizontal direction and is shown left-to-right. (a) $d_{CCV} = d_V$, (b) $d_{CCV} = d_V/2$, and (c) $d_{CCV} = d_V/3$.

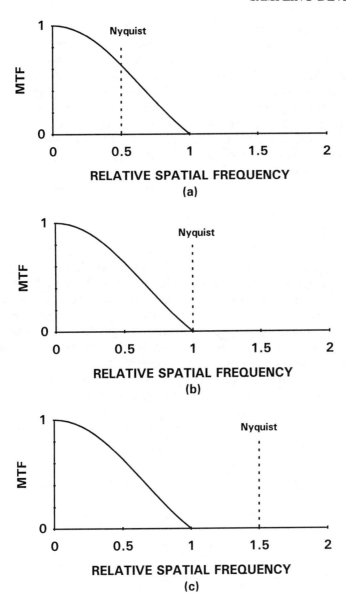

Figure 5-20. Vertical MTF$_{DETECTOR}$ as a function of v_i/v_{iD}. (a), (b), and (c) apply to the arrays shown in Figures 5-19a, 5-19b, and 5-19c, respectively. As the spacing, d_{CCV}, decreases, the Nyquist frequency increases.

120 SAMPLING, ALIASING, and DATA FIDELITY

The vertical scene components are reproduced with higher fidelity as center-to-center spacing decreases. However, if the system output conforms to a video standard, then the number of vertical detectors is limited by that standard. For example, RS 170 limits the number to approximately 480. Here, as the center-to-center spacing decreases, the vertical field-of-view decreases. This creates a tradeoff between image fidelity and vertical field-of-view. The center-to-center spacing selected depends on the final interpreter of image quality and the task. Staring arrays, which produce excellent imagery for most applications, provide only one sample per PAS (pixel-angular-subtense). This suggests that the configuration in Figure 5-19b or 5-19c should be used when (a) system MTF is critical or (b) edge location must be accurately located (minimize phasing effects).

The scan direction can be digitized at any rate by an ADC. Each horizontal sample represents an effective detector location. The time it takes for the line-of-sight to move one detector width is called the detector dwell time. In Figure 5-21a, the detector line-of-sight has moved one detector width during the sampling period. This represents one sample-per-dwell and provides the same output as a contiguous staring array. In Figure 5-21b, the line-of-sight moves only one-half of a detector width during a sample period and this provides two samples-per-dwell. Higher sampling rates are possible but they increase hardware complexity. Because of the misleading representation of the detector MTF (Figure 5-6, page 105) some analysts say that two samples-per-dwell satisfies the Nyquist criterion. As a result, it is the most prevalent design.

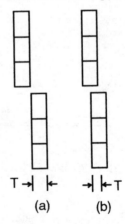

Figure 5-21. The upper figures represent the effective detector locations at time t and the lower figures represent the effective locations at time t + T. (a) One sample per detector dwell and (b) two samples per dwell. The sampling frequency is $f_s = 1/T$.

SAMPLING DEVICES 121

Similar considerations apply to the number of samples in the scan direction. If the output conforms to a video standard, the number of displayed samples may be limited by the video bandwidth. It is generally desirable to create square samples. That is, the effective center-to-center spacing should be the same in the horizontal and vertical directions. This simplifies image processing and display requirements. Again, because of the misleading representation of the detector MTF some analysts say that two samples-per-dwell satisfies the Nyquist criterion. As a result, it is the also most prevalent design in the scan direction.

D'Agostino et. al. performed[12] several perception tests in which trained observers were required to identify the correct vehicle and aspect angle. These vehicles were part of a computer data set of typical military vehicles (tanks, trucks, and jeeps). The sampling lattice and effective detector size were varied which changed the number of samples per pixel. Their results are:

a. When the sampling rate was greater than two per pixel, there was little additional improvement in performance.
b. One sample per pixel resulted in good performance considering the degree of aliasing assumed present in the imagery.

Higher sampling rates (more than four samples per pixel) reduced phasing effects and provided a cosmetically pleasing image. However, the observers were not asked to comment on the image quality. Only if they could correctly identify the target. The extension of their results to any other vehicle, ship, or aircraft is purely hypothetical. These results are consistent with commercial television. We like our television image even though it is one sample per pixel.

5.3.7. FINITE ARRAYS

A finite array may be considered as an infinite array with a window function that just encompasses the required number of detectors (Figure 5-22). The horizontal array size is $d_{ARRAY-H} = (N_H - 1)d_{CCH} + d_H$, where N_H is the number of detectors. In space the response is,

$$I_{FINITE}(x) = \left[I(x) * rect\left(\frac{x}{d_H}\right) \right] \cdot \left[comb\left(\frac{x}{d_{CCH}}\right) \cdot rect\left(\frac{x}{d_{ARRAY-H}}\right) \right]. \quad (5\text{-}13)$$

The amplitude spectrum becomes

$$I_{FINITE}(u_i) = \left[I(u_i) | sinc(d_H u_i) | \right] * \left[comb\left(\frac{u_i}{u_{iS}}\right) * | sinc(d_{ARRAY-H} u_i) | \right]. \quad (5\text{-}14)$$

122 SAMPLING, ALIASING, and DATA FIDELITY

For an infinite array, $d_{ARRAY-H} \to \infty$, $\text{sinc}(d_{ARRAY-H}u_i) \to 1$ and the convolution in the bracket just returns $\text{comb}(u_i/u_{is})$. This becomes the same expression as Equation 5-12. For one detector, $d_{ARRAY-H} \to d_H$ and $d_{CCH} \to d_H$. Then the zeros of $\text{sinc}(d_H u_i)$ fall on $\text{comb}(u_i/u_{is})$. This leaves just one value $\delta(f)$ and the convolution returns the analog detector response given by Equation 5-8 (page 103). Because $d_{ARRAY-H}$ is usually much larger than d_H, $\text{sinc}(d_{ARRAY-H}u_i)$ is usually neglected. Replacing $I_{FINITE}(u_i)$ with $I_{INFINITE}(u_i)$ is an idealization that is not physically realizable but it simplifies the mathematics. Because the reconstruction filter only passes the base band and sometimes part of the first few replicated bands, neglecting $\text{sinc}(d_{ARRAY-H}u_i)$ will not significantly affect the reconstructed signal appearance.

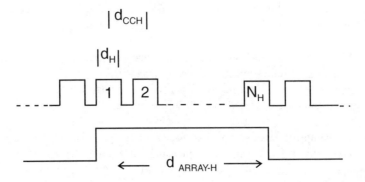

Figure 5-22. A window function creates a finite array.

5.3.8. OPTICAL PREFILTERING

Because aliasing occurs at the detector, the signal must be optically band-limited to the Nyquist frequency. Optical band-limiting can be achieved by using small diameter optics, defocusing, blurring the image, or by inserting a birefringent crystal between the lens and array. The birefringent crystal increases[4] the effective detector size but does not change detector locations. Therefore, the Nyquist frequency remains constant but u_{iN}/u_{iD} increases. This crystal is found in almost all single-chip color cameras.

Unfortunately, these approaches also degrade the MTF (reduce image sharpness) in the base band and typically are considered undesirable for scientific applications. For medical applications, where aliased imagery may be misinterpreted as an abnormality, reduced aliasing usually out weighs any MTF degradation. In a series of experiments performed by Barbe and Campana[13] with

SAMPLING DEVICES 123

monochrome CCDs, they conclude *the required prefiltering to reduce the response beyond Nyquist appears to do more harm than good.* Image quality is not significantly affected when modest aliasing is present. However, color shifts are obvious. For single-chip color assemblies, the prefilter is required.

Aliasing in the vertical direction can be reduced by summing alternate detector outputs (field integration) in a two-field camera system.[14] Here, the effective detector area is $2d_V$ and the cutoff is $1/2d_V$ (Figure 5-23). This reduces aliasing but also reduces the vertical MTF and, therefore, reduces image sharpness in the vertical direction. Based on the misleading detector representation (Figure 5-6, page 105), it is said that alternate row summing *eliminates* aliasing. Horizontal sampling still produces aliasing.

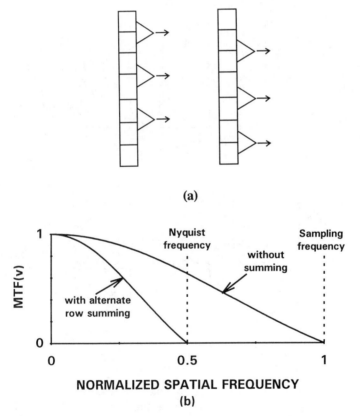

Figure 5-23. Alternate detector summing can reduce aliasing in the vertical direction. (a) Detector outputs and (b) vertical MTF with and without summing normalized to v_i/v_{iD}.

Note that arrays are often specified by the pixel size. With 100% fill-factor arrays, the pixel size is equal to the detector size. With finite fill-factor arrays, the photosensitive detector dimensions are less than the pixel dimensions. Microlenses[15] increase the effective detector size but do not affect the Nyquist frequency (Figure 5-24). By increasing the effective detector area, the detector cutoff decreases and u_{iN}/u_{iD} increases.

In color filter arrays, the pitch of like-color sensitive detectors may be larger than the individual detector pitch. This creates a low fill-factor for each color and potentially increases aliasing. The optical anti-alias filter, which consists of one or more birefringent crystals, is placed between the lens assembly and the detector array. Birefringent crystals break a beam into two components. Rays that would have missed the detector are refracted onto the detector. This makes the detector appear optically larger.[4]

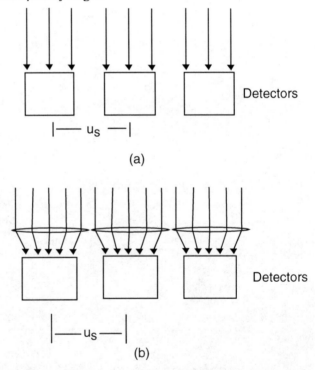

Figure 5-24. Optical effect of a microlens assembly. (a) With no microlens, a significant amount of photon flux is not detected. (b) The microlens assembly can image nearly all the flux onto the detector. Microlenses do not affect the sampling frequency. They make the detectors appear optically larger.

5.3.9. NONRECTANGULAR SAMPLING

Other sampling configurations may be more efficient[16,17]. Both atmospheric and optical spatial frequency responses tend to be rotationally symmetrical. Similarly, natural imagery appears to be symmetric. Therefore, the two-dimensional frequency spectrum reaching the detector plane is probably rotationally symmetric. A rectangular sampling lattice replicates these frequencies as illustrated in Figure 4-13a (page 94). The spaces between the circles offer no information. With hexagonal sampling, the individual frequencies are brought together in a "closest packing" configuration (Figure 5-25b).

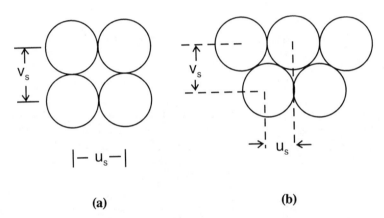

(a)   (b)

Figure 5-25. Most imagery tends to be rotationally symmetric. (a) Two-dimensional frequency spectrum with rectangular sampling and (b) spectrum with hexagonal sampling (after Reference 17). With hexagonal sampling, the vertical and horizontal sampling frequencies are different.

The hexagonal array offers two advantages over a square sampling grid. If the sampling frequencies are kept approximately constant, then the detectors can be made approximately 27% larger.[17] This increases the system responsivity. For detectors with the same center-to-center spacing, the horizontal sampling frequency increases by a factor of 2 and the vertical sampling frequency increases by $2/\sqrt{3}$ (Figure 5-26b). Hexagonal sampling provides[8,18] better MTF (higher MTF) than a comparable square sampling lattice. However, the MTFs are no longer separable[1] and this complicates analysis. The disadvantage is that data processing and data display are based on rectangular arrays. This requires resampling the datels from a hexagonal lattice into a rectangular lattice.

126 SAMPLING, ALIASING, and DATA FIDELITY

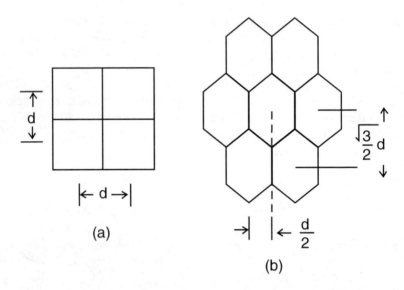

Figure 5-26. (a) Detectors in a square format and (b) detectors in a hexagonal configuration. The horizontal and vertical sampling frequencies are $2/d$ and $2/\sqrt{3}d$ cycles/mm, respectively. Rotating the hexagonal array by 90° interchanges the sampling frequencies.

5.3.10. FREQUENCY AXIS NORMALIZATION

When discussing sampling theory in Chapter 4, it was convenient to normalize the frequency response to the sampling frequency. With detectors, the MTFs were normalized to the detector cutoffs. Different normalizations make the MTFs *appear* different. Figure 5-27 illustrates the same MTFs plotted in Figure 5-11 (page 111) but with the frequency axis normalized to the array sampling frequency. To see the actual response, the MTF should be plotted in actual units (Figure 5-28) rather than normalized units.

SAMPLING DEVICES 127

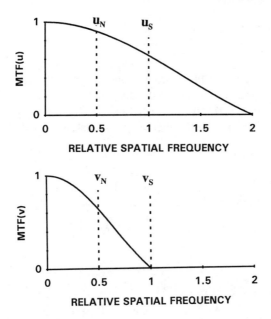

Figure 5-27. MTFs of Figure 5-11 (page 111) normalized to the array sampling frequency.

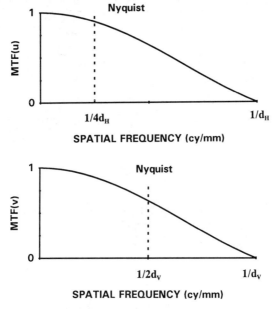

Figure 5-28. MTFs plotted as a function of u_i and v_i.

5.4. FRAME GRABBERS

For many applications, the camera output is an analog signal that is then re-sampled by a frame grabber (frame capture device). Frame grabbers are usually designed to accept standardized video signals (Figure 5-29). The encoded video signal is demodulated and then an ADC digitizes the signal. The conversion clock rate determines the number of datels created per line.

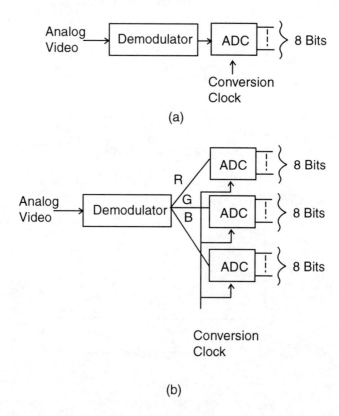

Figure 5-29. Frame grabbers. (a) Monochrome output and (b) color output. 8-bit ADCs are the most popular. 24 bits of color represents 2^{24} or 16.8 million color combinations. The HVS cannot discern all of these combinations.

Given the bandwidth and number of vertical lines, the number of datels is standardized. With detector arrays that match the standard, the number of datels will equal the number of pixels. However, the only requirement is that the camera *output* conform to the video standard. There is no *a priori* relationship between the number of frame grabber datels and number of detector pixels. Consider an array with 240 detectors in the vertical direction. With duplication, the camera output can match the NTSC standard. Similarly, the horizontal bandwidth is independent of the number of horizontal detectors. The Nyquist frequency associated with the frame grabber may be different than the array Nyquist frequency.[19] The image analyst must be made aware of this relationship.

If the signal is band-limited (as specified by the video standard) and the frame grabber clock frequency is twice that value, no internal anti-alias filter is required. Therefore, most frame grabbers do not have an internal anti-alias filter. Nonstandard signals with larger bandwidth can be aliased. Even if an anti-alias filter exists, it cannot remove aliasing that occurred at the detector. For end-to-end system analysis, the frame grabber Nyquist frequency must be added to the MTF curves. Additional aliasing by the frame grabber must also be considered. This is further discussed in Section 6.6., *Frame Grabbers*.

5.5. REFERENCES

1. O. Hadar, A. Dogariu, and G. D. Boreman, "Angular Dependence of Sampling MTF," *Applied Optics*, Vol. 36, pp. 7210-7216 (1997).
2. G. C. Holst, *CCD Arrays, Cameras, and Displays*, pp. 58, JCD Publishing, Winter Park, FL (1996).
3. G. C. Holst, *CCD Arrays, Cameras, and Displays*, pp. 59, JCD Publishing, Winter Park, FL (1996).
4. G. C. Holst, *CCD Arrays, Cameras, and Displays*, pp. 70-73, JCD Publishing, Winter Park, FL (1996).
5. J. E. Greivenkamp, "Color Dependent Optical Prefilter for the Suppression of Aliasing Artifacts," *Applied Optics*, Vol. 29(5), pp. 676-684 (1990).
6. G. C. Holst, *Electro-Optical Imaging System Performance*, pp. 96-98, JCD Publishing, Winter Park, FL (1995).
7. K. J. Barnard, E. A. Watson, and P. F. McManamon, "Nonmechanical Microscanning Using Optical Space-fed Phased Arrays," *Optical Engineering*, Vol. 33(9), pp. 3063-3071 (1994).
8. K. J. Barnard and E. A. Watson, "Effects of Image Noise on Submicroscan Interpolation," *Optical Engineering*, Vol. 34(11), pp. 3165-3173 (1995).
9. K. M. Hock, "Effect of Oversampling in Pixel Arrays," *Optical Engineering*, Vol. 34(5), pp. 1281-1288 (1995).
10. J. C. Gillette, T. M. Stadtmiller, and R. C. Hardie, "Aliasing Reduction in Staring Infrared Imagers Utilizing Subpixel Techniques," *Optical Engineering*, Vol. 34(11), pp. 3130-3137 (1995).
11. A. Friedenberg, "Microscan in Infrared Staring Systems," *Optical Engineering*, Vol. 36(6), pp. 1745-1749 (1997).

12. J. D'Agostino, M. Friedman, R. LaFollette, and M. Crenshaw, "An Experimental Study of the Effects of Sampling on FLIR Performance," in *Proceedings of the IRIS Specialty Group on Passive Sensors*, Infrared Information Analysis Center, Ann Arbor Mich. (1990).
13. D. F. Barbe and S. B. Campana, "Imaging Arrays Using the Charge-Coupled Concept," in *Image Pickup and Display, Volume 3*, B. Kazan, ed., pp. 245-253, Academic Press (1977).
14. G. C. Holst, *CCD Arrays, Cameras, and Displays*, page 63, JCD Publishing, Winter Park, FL (1996).
15. G. C. Holst, *CCD Arrays, Cameras, and Displays*, pp. 85-86, JCD Publishing, Winter Park, FL (1996).
16. J. Burton, K. Miller, and S. Park, "Fidelity Metrics for Hexagonally Sampled Digital Imaging Systems," *Journal of Imaging Science and Technology*, Vol. 17(6), pp. 279-283 (1991).
17. R. Legault, "The Aliasing Problems in Two-Dimensional Sampled Imagery," in *Perception of Displayed Information*, L. M. Biberman, ed., pp. 305-307, Plenum Press, New York (1973).
18. K. J. Barnard and G. D. Boreman, "Modulation Transfer Function of Hexagonal Staring Focal Plane Arrays," *Optical Engineering*, Vol. 30(12), pp. 1915-1919 (1991).
19. G. D. Boreman, "Fourier Spectrum Techniques for Characterization of Spatial Noise in Imaging Systems," *Optical Engineering*, Vol. 26(10), pp. 985-991 (1987).

6
RESAMPLING

Implicit in the c/d/c/d/c system (Figure 1-17, page 24) is that all the devices are matched. That is, the number of "-els" (scenels, pixels, datels, and disels) are identical. Resampling is necessary when the "-els" are different or when an "-el" is not at a desired location.

Resampling includes[1-4] both decimation and interpolation. With decimation, the sampling frequency is decreased and samples are discarded. Interpolation increases the sampling frequency and samples are added. It creates samples at locations other than where the original samples were taken. Interpolation does not increase the system resolution because no new frequencies are added. No additional detail is available. But decimation can reduce resolution. By using interpolation and decimation in tandem, any frequency (rational value) can be created.

Localized resampling alters only part of the image. For example, geometric correction[5] has been used for years by the remote sensing community. Corrections exist for altitude variations, attitude, scan-skew, velocity changes, earth rotation, map projection, panorama, and perspective. The same operations are now used in all computer drawing programs that allow rotation and stretching of any portion of the image in any direction. This is currently called image warping.[6-7]

To maintain resolution, Hewlett Packard uses resolution enhancement technology (RET®) in their LaserJet® printers. RET employs over 200 rules that are applied to a 49-pixel window to improve the printed quality. These rules, which are forms of localized resampling, are nonlinear operators and can only be analyzed on a case-by-case basis. Kang[8] describes various area and localized resampling techniques.

With electronic circuitry, the time between samples changes. This can be performed in near real-time. With electronic imaging systems, the data reside in a computer's memory and the size of the array is manipulated before reading the array. The array size expands for interpolation and contracts with decimation. Interpolation adds more datels for a fixed field-of-view and the sampling rate increases in terms of datels per field-of-view. Because data array elements are typically mapped one-to-one onto a display, the displayed image size changes. The success of resampling can only be ascertained after reconstruction (discussed in Chapter 7, *Reconstruction*).

A fundamental difference exists between electronic signal and image resampling. Electronic circuits are causal. That is, an output cannot occur before an input. The next sample can only be estimated from the preceding samples. With images, the interpolation algorithm can use samples on either side of the desired sample. However, digital circuits may employ time delay so that datels on either side of the desired sample are considered.

This text describes global resampling uniformly applied to the entire data set. Here, resampling can be analyzed in both the time/space and frequency domains. The symbols used in this book are summarized in the *Symbol List* (page xiii) which appears after the *Table of Contents*.

6.1. EXAMPLES

Resampling occurs frequently when two different devices are linked together. When the devices use common communication formats, the resampling operation is often transparent to the user. For example most cameras and monitors conform to standard video formats (e.g., EIA 170, NTSC, PAL, or SECAM). This ensures that the imagery can be displayed without any additional interfacing. Although the number of pixels and disels is typically matched to the video standard, they may be different. For communication links, frequency conversion is required to transmit signals over allocated channels. Because most processing is digital, these systems are called[4] *multirate digital signal processing systems*. Mapping "-els" is sometimes called resolution conversion. This mapping ensures that the resolution is maintained as the data flow from one device to the next.

In document transmittal, maintaining the same sized image across various devices is important. Fax machines must be able to send and print identically sized images. If the sending document scanner operates at 204 dots-per-inch (dpi) horizontally and the receiving fax machine prints at 300 dpi, the image must be appropriately scaled by the receiver. Similarly, flat bed scanners may operate at 300 dpi but the computer monitor may offer only 87 dpi (equivalent to 0.28 mm dot pitch).

Remapping operations include zoom, rotation, and multiple image registration to a common coordinate system. The last operation is often used in remote sensing to overlay imagery collected from a variety of sensors. When applied to a specific application, this procedure is more popularly called sensor fusion.

Figure 6-1 illustrates two-dimensional resampling. The original and desired sampling lattices are illustrated in Figures 6-1a and 6-1b, respectively. The algorithm that estimates a sample may use the intensity values of the four nearest neighbors (Figure 6-1c). Figure 6-1 provides the conceptual relationship between the two sampling lattices. The data arrays change size according to the resampling operation.

Figure 6-2 illustrates the spatial sampling of a rectangular target by contiguous square detectors. The digitized output of each detector is placed into a data array. Each array element is read out to a disel using a zero-order reconstruction filter. Here, there is a one-to-one mapping from pixels to datels and from datels to the disels. Phasing effects (discussed in Section 8.1.) further complicate the following discussion.

In Figure 6-3, the detector aspect ratio is 2:1. If the data array is mapped 1:1 onto the display, then the image will be distorted in the horizontal direction. Without *a priori* knowledge of the detector sampling lattice, an image analyst may assume that the object is only two samples wide and that the vertical and horizontal extent are equal. Because the detectors have an aspect ratio of 2:1, the interpolation algorithm must increase the number of horizontal samples by two to create an apparent square sampling lattice. After interpolation, the image analyst must remember that resolution depends on the original data set and not the interpolated data set. Note that the data array in Figure 6-3 contains 4 × 4 datels and after interpolation the new array will contain 8 × 4 datels.

Figures 6-2 and 6-3 illustrated a zero-order reconstruction filter which created a blocky image. Zoom is a very popular interpolation operation and the reconstruction filter affects its appearance. Therefore, zoom is discussed with other reconstruction filters in Chapter 7, *Reconstruction*.

134 SAMPLING, ALIASING, and DATA FIDELITY

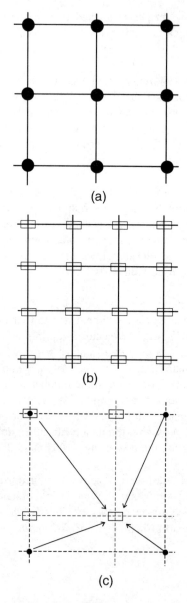

Figure 6-1. Conceptual relationship with two-dimensional resampling. (a) Original sampling lattice (dots), (b) desired sampling lattice (squares), and (c) lattices superimposed. Here, the sample value is estimated from the four nearest neighbors.

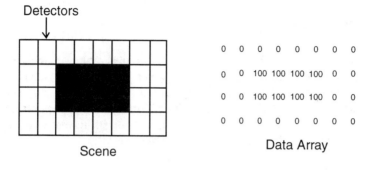
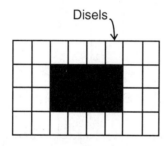

Figure 6-2. Two-dimensional sampling by contiguous square detectors. The output of each detector is placed in a data array mapped one-to-one onto the display. The data array values are digital numbers.

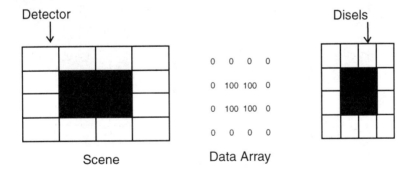

Figure 6-3. Detector arrays with non-square pixels create unequal sampling lattices in the horizontal and vertical directions. If the data array is mapped one-to-one onto the display, the image is distorted. Phasing effects further alter the displayed image width, height, and intensity.

136 SAMPLING, ALIASING, and DATA FIDELITY

6.2. DECIMATION

Decimation is the uniform removal of data and thereby lowers the sampling frequency. The new sampling frequency is $f_{S2} = f_{S1}/n_S$, where f_{S1} is the original frequency. In Figure 6-4, the sampling frequency is halved by discarding every other pulse ($n_S = 2$). If the original signal is band-limited to $f_{S1}/(2n_S)$, no aliasing will occur (Figure 6-5). If this is not true, an anti-alias filter must be inserted before the down sampling operation to avoid aliasing.

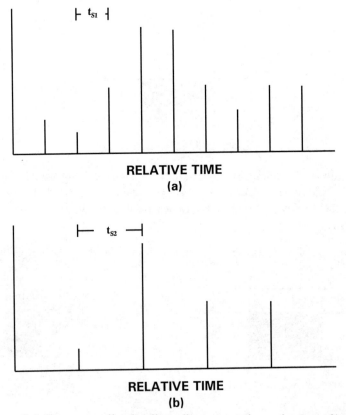

Figure 6-4. Down sampling by discarding every other pulse ($n_S = 2$). (a) The original signal and (b) the signal after down sampling. The time between pulses has increased to $t_{S2} = 2t_{S1}$. Similarly, the frequency has decreased to $f_{S2} = f_{S1}/2$.

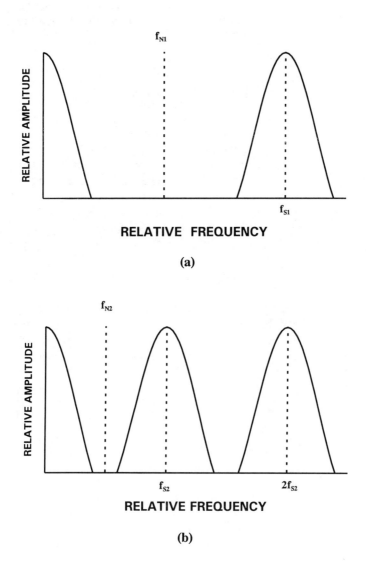

Figure 6-5. (a) The original signal and (b) the down sampled frequency spectrum when $n_S = 2$. If the signal contains frequencies greater than f_{N2}, aliasing will occur. $f_{N2} = f_{S1}/(2n_S)$. Only positive frequencies are illustrated.

On the surface, it seems that sample averaging is appropriate when down sampling. However, its effectiveness depends on the frequency spectrum and the reconstruction filter used. When N_{AVE} samples are averaged, the MTF is[9]

$$MTF_{AVE}(f) = \left| \frac{\sin\left(N_{AVE}\pi \frac{f}{f_{S1}}\right)}{N_{AVE}\sin\left(\pi \frac{f}{f_{S1}}\right)} \right| . \qquad (6\text{-}1)$$

The first zero of this function occurs at f_{S1}/N_{AVE}. When down sampling, N_{AVE} should be $2n_S$ so that attenuation occurs above f_{N2}. Figure 6-6 illustrates the averaging effects for $n_S = 2$ ($N_{AVE} = 4$). Averaging filters are not efficient low-pass filters. They have considerable response beyond f_{N2} and this leads to aliasing. By attenuating the in-band frequencies (below f_{N2}), the reconstructed signal is modified. Digital low-pass filters are further discussed in Section 9.5.1., *Low-pass Filters*.

Decimation in imagery is the same as decimation with electronic signals: samples are discarded. Discarding samples is equivalent to reducing the fill-factor. The aliased spectrum with 100% fill-factor was illustrated in Figure 5-16 (page 115). If $n_S = 2$, the resultant spectrum is identical to that obtained with a 50% fill-factor (Figure 5-11b, page 111). Decimation reduces the number of datels. With one-to-one mapping from datels to disels, the displayed image is physically smaller. That is, decimation minifies the image.

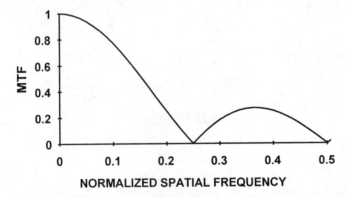

Figure 6-6. MTF associated with averaging four samples. The frequency axis is normalized to f_{S1} ($f_{S1} = 1$). Digital filter response is symmetrical about the Nyquist frequency ($f_{N1} = 0.5$).

6.3. INTERPOLATION

Interpolation is the estimate of a signal at locations other than where it was sampled. Although the samples were independently collected, the signal is assumed to be continuous. With integer interpolation, the new sampling frequency is $f_{S2} = m_S f_{S1}$. For illustration, consider a sampling system that creates three times more samples ($m_S = 3$). Then $f_{S2} = 3f_{S1}$ and $t_{S2} = t_{S1}/3$. At each new data location, a zero is inserted (Figure 6-7b). The interpolation algorithm estimates the data value at each new location (Figure 6-7c).

Figure 6-7b can be interpreted as the result of multiplying two signals together: the original signal and a second with m_S more samples. For LSI systems, multiplication in the time domain is equivalent to convolution in the frequency domain. As illustrated in Figure 6-8b, the original frequency spectrum (Figure 6-8a) is convolved with the sampling lattice at $\pm n f_{S2}$. This creates additional spectra about $f_{S2}/3$ and $2f_{S2}/3$.

The new sampling frequency suggests that the ideal reconstruction filter should be unity up to f_{N2}. With the presence of the intermediate spectra, the reconstruction filter should be unity up to f_{N1} (Figure 6-9). Amplitudes of the intermediate spectra depend on the interpolation algorithm. Low-pass filtering can further reduce these amplitudes. The intermediate spectra centered on $\pm n f_{S2}/n_S$ are created by the interpolation and would not exist if the signal was originally sampled at f_{S2}.

140 SAMPLING, ALIASING, and DATA FIDELITY

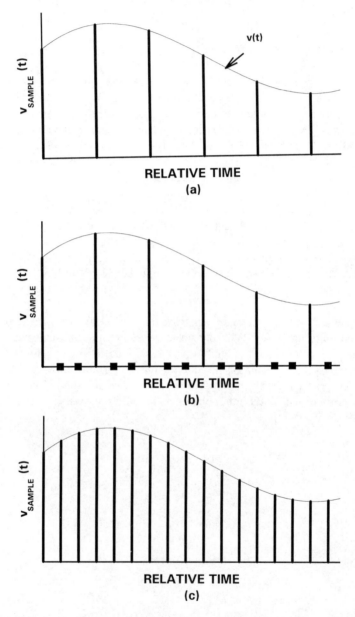

Figure 6-7. (a) Original sampled signal. (b) Insertion of zeros to create an apparent higher sampling rate. (c) Original signal with estimated values.

RESAMPLING 141

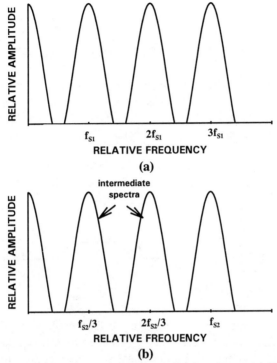

Figure 6-8. Frequency conversion and amplitude spectra. (a) Frequency spectra of a band-limited signal sampled at f_{S1}. (b) By adding zeros, the spectrum is replicated about $\pm nf_{S2}/n_S$. The interpolation algorithm may remove the intermediate spectra.

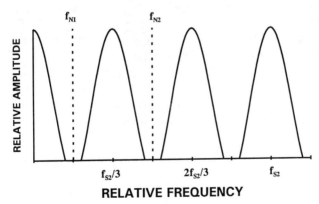

Figure 6-9. If the reconstruction filter cutoff is based on f_{N2}, new frequencies, not present in the original signal, may be reconstructed.

142 SAMPLING, ALIASING, and DATA FIDELITY

6.4. FREQUENCY SHIFT (RATIONAL VALUE)

It is easier to understand and analyze frequency shifts by artificially creating an intermediate stage that is the least common multiple of the input and output sampling rates. For example, if $f_{S2} = m_S f_{S1}/n_S$ then the sampling rate increases by m_S times and then decreases by n_S times (Figure 6-10). By increasing the sampling rate m_S times, both the input and output can be mapped to the same digital samples. Then the sampling rate is reduced n_S times by discarding the data points not needed. In practice, only the required data points are generated.

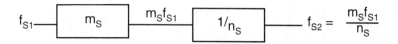

Figure 6-10. The fictitious intermediate step illustrates the frequency change. Conceptually, the frequency is increased m_S times and then decreases n_S times. In real circuits, this intermediate step does not exist and one algorithm performs the complete operation. An anti-alias filter may be necessary before the down sampling operation.

For illustration, let $f_{S2} = 3f_{S1}/2$ and then the figures of Sections 6-2 and 6-3 can be used. The artificial intermediate sampling stage has three times more samples (Figure 6-11). Then every other pulse is ignored. The frequency spectrum first expands as shown in Figure 6-8 and then compresses (see Figure 6-5) to provide the final spectrum shown in Figure 6-12. Low-pass filtering may be necessary to prevent aliasing of the intermediate spectra during the down sampling operation. If the remnants centered at $f_{S2}/3$ and $2f_{S2}/3$ are removed by a low-pass digital filter, the reconstruction filter cutoff can be f_{N2}. Otherwise, the reconstruction filter cutoff should be f_{N1}/n_S.

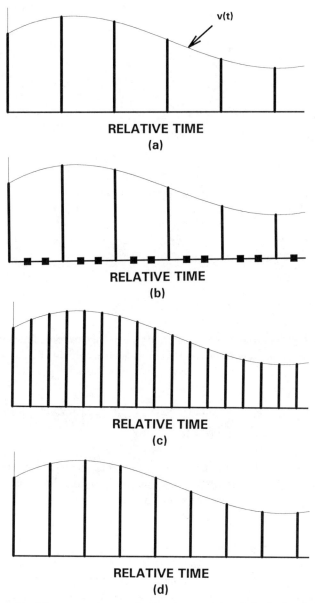

Figure 6-11. The steps (top to bottom) are: (a) original signal to be frequency shifted, (b) zeros are inserted to increase the sampling rate, (c) the interpolation algorithm estimates the values, and (d) every other value is discarded to create a final frequency of $3/2f_{S1}$.

144 SAMPLING, ALIASING, and DATA FIDELITY

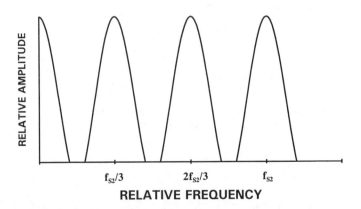

Figure 6-12. The final amplitude spectrum after the 3/2 frequency shift. The intermediate spectra are typically removed with a digital low-pass filter. Only positive frequencies are illustrated.

6.5. INTERPOLATION ALGORITHMS

Assuming a continuous, band-limited signal, the interpolation algorithm will not create any new frequency not present in the original signal. The interpolated values are

$$v'_{SAMPLE}(kt_{S2}) = \sum_{n=-\infty}^{\infty} v_{SAMPLE}(nt_{S1}) \, r_i(kt_{S2}) \, , \qquad (6\text{-}2)$$

where k and n are integers based on a common, arbitrarily selected, origin, $r_i(kt_{S2})$ is the interpolation algorithm, $v_{SAMPLE}(nt_{S1})$ is the original sampled signal, and $v'_{SAMPLE}(kt_{S2})$ is the expected sampled signal when the continuous signal v(t) is sampled every t_{S2} sec.

Because the original signal contains frequencies up to f_H, the interpolation algorithm should pass these frequencies without any attenuation. The ideal interpolation algorithm has unity amplitude up to the Nyquist frequency, f_{N1}, and then zero. This provides a sinc function interpolation to the newly generated data points:

$$v'_{SAMPLE}(kt_{S2}) = \sum_{n=-\infty}^{\infty} v_{SAMPLE}(nt_{S1}) \, sinc\left(\frac{nt_{S1} - kt_{S2}}{t_{S1}}\right) . \qquad (6\text{-}3)$$

RESAMPLING 145

The sinc function provides the multipliers for the original signal to obtain the new data set (Figure 6-13). It is not possible to evaluate the equation because it requires an infinite number of data points (the sum goes from $-\infty$ to ∞). That is, the ideal filter is not physically realizable.

The sinc function has negative values and this leads to the possibility that the estimated value may also be negative. While electronic signals may be both positive and negative, negative values have no meaning with imagery. Intensities are only positive. A normalization dilemma also exists. For example, if the signal is represented by 8 bits (DN ranges from 0 to 255), the negative values must be forced to zero.

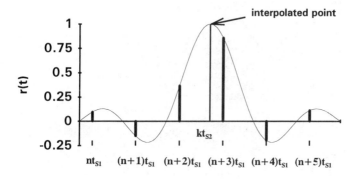

Figure 6-13. Ideal interpolator. Each data point, $v_{SAMPLE}(nt_{S1})$, is multiplied by sinc to create $v'_{SAMPLE}(kt_{S2})$.

Many interpolation filters are modified sinc functions. The sinc function is windowed so that the values are zero outside a specified range (Figure 6-14). This filter is physically realizable but may still have negative components. To minimize computational complexity, the window often encompasses only a few points (less than 10). Interpolator design is a tradeoff between amplitude response in the pass band (up to f_H) and the amplitude at the replicated frequencies ($f > f_{S1} - f_H$). As f_{S1} decreases to $2f_H$, algorithm complexity increases.

Figures 6-13 and 6-14 illustrate a filter with symmetric response about zero. This means that sample values on either side of the desired sample are used. This is only possible with electronic circuits that include a time delay. By restricting the algorithm to N samples, the time delay is $Nt_{S1}/2$ sec.

146 SAMPLING, ALIASING, and DATA FIDELITY

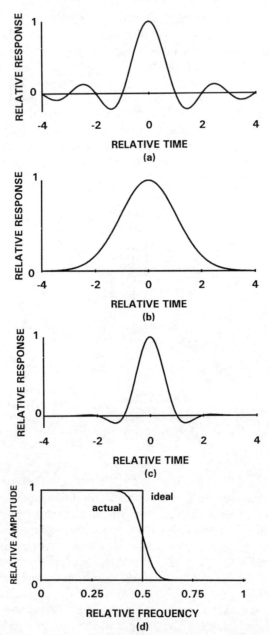

Figure 6-14. Most interpolation algorithms are windowed sinc functions. (a) sinc function, (b) window, (c) interpolator, and (d) frequency spectra of a real algorithm.

With two-dimensional interpolation,

$$f(k_1\Delta x', k_2\Delta y') = \sum_{m_{s1}=-\infty}^{\infty} \sum_{m_{s2}=-\infty}^{\infty} f(m_{s1}\Delta x, m_{s2}\Delta y) \, r_i(k_1\Delta x', k_2\Delta y') , \quad (6\text{-}4)$$

where Δx and Δy are the data increments of the original image, $\Delta x'$ and $\Delta y'$ are the data increments of the interpolated image, and m_{s1} and m_{s2} may be different. The equation is evaluated for each value of $k_1\Delta x'$ and $k_2\Delta y'$.

If the original array is N × M, then the new array is m_{s1}N × m_{s2}M. Here, the summations are from 0 to m_{s1} and 0 to m_{s2}. If $m_{s1} = m_{s2}$, it is simply called zoom or electronic zoom. This is further described in Section 7.4., *Electronic Zoom*. For convenience, the interpolator is considered separable:

$$r_i(k_1\Delta x', k_2\Delta y) = r_i(k_1\Delta x') \, r_i(k_2\Delta y') . \quad (6\text{-}5)$$

Equations 6-2 and 6-4 describe interpolation in terms of time or space. Digital algorithms operate on an data array where elements are described by indices. The interpolated values are

$$x'(k) = \sum_{n=0}^{N-1} x(nm_s) \, r_i(k) , \quad (6\text{-}6)$$

where nm_s is the location of the original data set in the new array, the maximum value of n and k are N - 1 and K - 1, respectively, and K - 1 = m_s(N - 1) because k = nm_s.

Spline interpolation functions are generated by successive convolutions of a rectangular response. The simplest interpolator is the zero-order algorithm. Each new sample is just a replication of its nearest neighbor. With the first-order algorithm, the original signal is assumed to be linear between the samples. Higher order algorithms allow for signal curvature. The algorithm order, m, is equal to the number[10] of convolutions. In addition, for the mth-order algorithm, the first m-1 derivatives are continuous. The 3rd-order, or cubic spline, is the most popular. It uses only four samples: two on either side of the desired sample. Table 6-1 provides the continuous representation of the spline functions.

148 SAMPLING, ALIASING, and DATA FIDELITY

Table 6-1
ONE-DIMENSIONAL INTERPOLATION ALGORITHMS

ORDER	SPATIAL RESPONSE	FREQUENCY RESPONSE
Ideal	$r_{IDEAL}(x) = \text{sinc}(x)$	$R_{IDEAL}(u) = \text{rect}(u)$
Zero	$r_0(x) = \text{rect}(x)$ (pulse)	$R_0(u) = \text{sinc}(u)$
1st	$r_1(x) = r_0(x) * r_0(x)$ (triangle or linear)	$R_1(u) = R_0(u)R_0(u)$ $R_1(u) = \text{sinc}^2(u)$
2nd	$r_2(x) = r_1(x) * r_0(x)$ (bell)	$R_2(u) = R_1(u)R_0(u)$ $R_2(u) = \text{sinc}^3(u)$
3rd	$r_3(x) = r_2(x) * r_0(x)$ (cubic B-spline)	$R_3(u) = R_2(u)R_0(u)$ $R_3(u) = \text{sinc}^4(u)$
–	–	–

The following describes filter response in general terms.[11] Specific equations can be found in Reference 6. The zero-order algorithm is (Figure 6-15) is

$$r_o(k) = 1 \quad \text{when} \quad |z| \leq 0.5 \tag{6-7}$$
$$= 0 \quad \text{elsewhere},$$

where

$$z = \frac{nm_s - k}{m_s}. \tag{6-8}$$

For a particular value where $k' = k$,

$$x'(k') = x(nm_s) \quad \text{when} \quad \left|\frac{nm_s - k'}{m_s}\right| \leq 0.5. \tag{6-9}$$

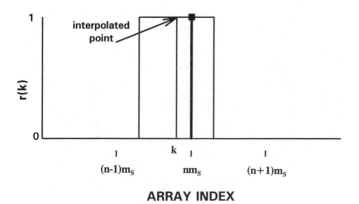

Figure 6-15. The rect(x) function simply replicates the nearest neighbor. The multiplicative constant is one. The array index applies to the interpolated image with values ranging from 0 to K - 1.

Figure 6-16 illustrates the linear interpolation algorithm:

$$r_1(k) = 1 - |z| \quad \text{when } |z| \leq 1$$
$$= 0 \quad \text{elsewhere .} \tag{6-10}$$

For a particular value where $k' = k$,

$$x'(k') = x(nm_s)\left(1 - \left|\frac{nm_s - k'}{m_s}\right|\right)$$
$$+ x((n+1)m_s)\left(1 - \left|\frac{(n+1)m_s - k'}{m_s}\right|\right) . \tag{6-11}$$

When used in two directions, it is called bilinear interpolation (Figure 6-17):

$$r_1(x,y) = tri(x)\,tri(y) . \tag{6-12}$$

150 SAMPLING, ALIASING, and DATA FIDELITY

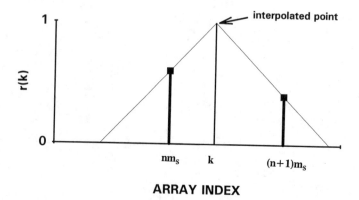

Figure 6-16. The tri(x) function weighs the values of the two nearest neighbors. The squares are the multiplicative constants. The array index applies to the interpolated image.

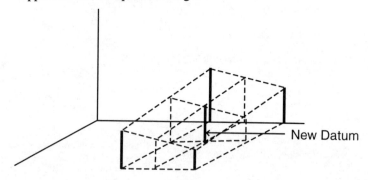

Figure 6-17. Bilinear interpolation. The surface connecting the four data points is not necessarily flat.

The 2nd-order interpolator is:

$$r_2(k) = \frac{3}{4} - z^2 \qquad \text{when } |z| \le \frac{1}{2}$$
$$= \frac{1}{2}\left(|z| - \frac{3}{2}\right)^2 \qquad \text{when } \frac{1}{2} \le |z| \le \frac{3}{2} \qquad (6\text{-}13)$$
$$= 0 \qquad \text{elsewhere .}$$

Figure 6-18 illustrates the cubic B-spline response which is described by:

$$r_3(k) = \frac{1}{2}|z|^3 - (z)^2 + \frac{2}{3} \quad \text{when } 0 \leq |z| \leq 1$$

$$= \frac{1}{6}(2 - |z|)^3 \quad \text{when } 1 \leq |z| \leq 2 \quad (6\text{-}14)$$

$$= 0 \quad \text{elsewhere .}$$

Cubic splines are often used because the 1st and 2nd derivatives are continuous. The parametric cubic convolution is

$$r_{PARA} = (\alpha + 2)|z|^3 - (\alpha + 3)|z|^2 + 1 \quad \text{when } 0 \leq |z| \leq 1$$

$$= \alpha|z|^3 - 5\alpha|z|^2 + 8\alpha|z| - 4\alpha \quad \text{when } 1 \leq |z| \leq 2 \quad (6\text{-}15)$$

$$= 0 \quad \text{elsewhere .}$$

As illustrated in Figure 6-19, when[12] $\alpha = -0.5$, the central lobe approximates the sinc function. The second lobes are positive and this avoids the negative values associated with a truncated sinc function (Figure 6-14c, page 146).

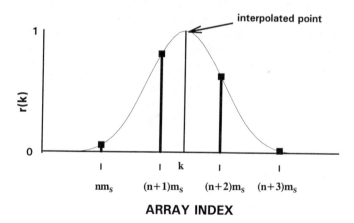

Figure 6-18. Cubic B-spline interpolation algorithm. The squares are the multiplicative constants. The array index applies to the interpolated image.

152 SAMPLING, ALIASING, and DATA FIDELITY

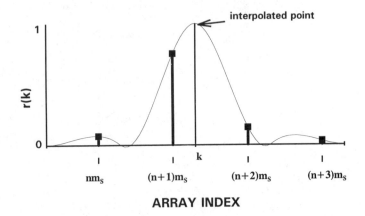

Figure 6-19. Parametric cubic interpolation algorithm. $\alpha = -0.5$. The squares are the multiplicative constants. The array index applies to the interpolated image.

The frequency response of the first four splines is illustrated in Figure 6-20. The frequency response of the cubic spline has positive values so that intensities remain positive. As the order increases, the amplitudes of frequencies above f_{N1} are attenuated so that the intermediate spectra are suppressed. However, this also attenuates in-band frequency amplitudes. In comparison, the ideal interpolator has unity response up to f_{N1}. From a mathematical viewpoint, the ideal filter is desirable. But from a practical viewpoint, it creates ringing at sharp edges (Gibbs phenomenon). Therefore, a good filter trades ringing with in-band attenuation. The parametric cubic convolution[12] is one filter that provides this tradeoff. Its frequency response is

$$R(u) = \frac{3\,sinc^2(u) - 3\,sinc(2u) + 6\alpha\,sinc^2(2u) - 4\alpha\,sinc(2u) - 2\alpha\,sinc(4u)}{(\pi u)^2}, \quad (6\text{-}16)$$

When $\alpha = -0.5$,

$$R(u) = \frac{3\,sinc^2(u) - sinc(2u) - 3\,sinc^2(2u) + sinc(4u)}{(\pi u)^2}. \quad (6\text{-}17)$$

As illustrated in Figure 6-21, this filter provides higher response in the pass band (compared to the 3rd-order spline) but at the expense of passing some frequencies above the Nyquist frequency.

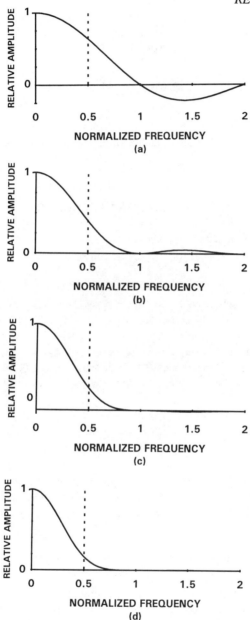

Figure 6-20. Frequency response of the various spline interpolators. (a) Zero-, (b) 1st-, (c) 2nd-, and (d) 3rd-order interpolators. The vertical dashed line indicates the Nyquist frequency and $f_S = 1$.

154 SAMPLING, ALIASING, and DATA FIDELITY

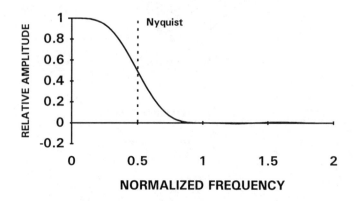

Figure 6-21. Frequency response of the parametric cubic interpolator when $\alpha = -0.5$.

With images, a whole frame of data is usually available. Interpolation can use data from its nearest neighbors both horizontally and vertically. The neighborhood is often limited to a few samples (typically 10 or less). Interpolation algorithms cannot operate at the edges of images because an insufficient number of elements exist. Asymmetric sampling occurs frequently and one-dimensional interpolation is employed to create an apparent square sampling lattice.

6.6. FRAME GRABBERS

Frame grabbers are designed to capture standard video signals. The nominal[13] video bandwidth of NTSC, PAL, and SECAM is 4.2, 5.5, and 6 MHz, respectively. According to the sampling theorem, the required number of samples per line is

$$N_{SAMPLE} = 2(active\ time\ line) \cdot (band\ width) . \qquad (6\text{-}18)$$

Table 6-2 lists the nominal number of horizontal samples per line. The quoted number varies in the literature and depends on the precise bandwidth and active line time selected by the author. The frame grabber's sampling frequency should be more than twice the allocated video bandwidth. Equivalently, the number of horizontal samples should be greater than that given in Table 6-2. The frame grabber's internal ADC samples the signal in the horizontal direction only.

The vertical sampling rate is dictated by the number of vertical lines specified by the video format. To maintain the aspect ratio (see Figures 6-1 and 6-2, pages 134-135), frame grabbers usually offer square disels. Because EIA 170/NTSC has an aspect ratio of 4:3 and the number of vertical lines is 480, most EIA 170-compatible frame grabbers provide 640 horizontal datels.

Table 6-2
NUMBER OF HORIZONTAL VIDEO ELEMENTS
Based on electrical bandwidth only

FORMAT	ALLOCATED BANDWIDTH (MHz)	ACTIVE LINE TIME (μs)	NUMBER of HORIZONTAL ELEMENTS
EIA 170/NTSC	4.2	52.09	438
PAL	5.5	51.7	569
SECAM	6.0	51.7	621

Generally, the detector array is also matched to the video standard. An imaging system with more detectors cannot provide any more *system* resolution if it conforms to the standard video bandwidth limitation.

Most frame grabbers do not have an anti-alias filter. Aliasing within the frame grabber depends on the sharpness of the reconstruction filter within the camera. Unfortunately, these filter characteristics are difficult to obtain. The scene is sampled by the detectors, digitized, and then converted into an analog signal. This analog signal contains scene components aliased by the detectors. If the analog signal is band-limited by an ideal reconstruction filter, then no frequency exists above f_N. If the frame grabber sampling frequency is greater than $2f_N$, then no additional aliasing will occur.

However, real filters pass some frequencies above f_N and these frequencies can be aliased by the frame grabber. Figure 6-22 illustrates a frame grabber whose sampling frequency that is 2.4 times the detector array Nyquist frequency. That is, the number of frame grabber datels is 1.2 times the number of detectors. This can easily occur in low-cost visible-system detector arrays. The amount of aliasing depends on the sharpness of the reconstruction filter used to create the analog signal and the frame grabber sampling frequency.

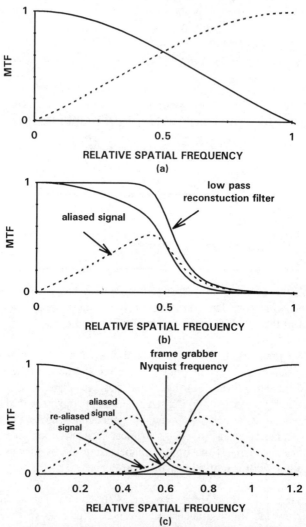

Figure 6-22. Resampling by a frame grabber. (a) Detector MTF and aliasing (showing first frequency replication only) created by a staring array with 100% fill-factor. The frequency axis is normalized to the array sampling frequency ($f_S = 1$). (b) Attenuated in-band and aliased signal after an 8th-order, low-pass, Butterworth filter. (c) Aliasing created by a frame grabber whose sampling frequency is 1.2 times the staring array sampling frequency.

6.7. REFERENCES

1. R. E. Crochiere and L. R. Rabiner, "Interpolation and Decimation of Digital Signals - A Tutorial Review," *Proceedings of the IEEE*, Vol. 69(3), pp. 300-331 (1981).
2. R. W. Schafer and L. R. Rabiner, "A Digital Signal Processing Approach to Interpolation," *Proceedings of the IEEE*, Vol. 61(6), pp. 692-702 (1973).
3. A. V. Oppenheim and R. W. Schafer, *Digital Signal Processing*, Chapter 5, Prentice Hall, Englewood Cliffs, NJ (1975).
4. J. G. Proakis and D. G. Manolakis, *Digital Signal Processing: Principles, Algorithms, and Applications*, 3rd Edition, pp. 782-846, Prentice-Hall, Upper Saddle, NJ (1996).
5. See, for example, R. Bernstein, "Image Geometry and Rectification," in *Manual of Remote Sensing*, 2nd edition, R. N. Colwell, ed., Chapter 21, American Society of Photogrammetry, Fall Church, VA (1983).
6. A. R. Weeks, Jr., *Fundamentals of Electronic Image Processing*, pp. 294-315, SPIE Optical Engineering Press (1996).
7. J. C. Russ, *The Image Handbook*, 2nd Edition, pp. 195-207, CRC Press, Boca Raton, FL (1994).
8. Henry Kang, *Color Technology for Electronic Imaging Devices*, pp. 177-207, SPIE Optical Engineering Press, Bellingham, WA (1997).
9. H. V. Kennedy, "Miscellaneous Modulation Transfer Function (MTF) Effects Relating to Sampling Summing," in *Infrared Imaging Systems: Design, Analysis, Modeling, and Testing*, G. C. Holst, ed., SPIE Proceedings Vol. 1488, pp. 165-176 (1991).
10. H. C. Andrews and C. L. Patterson, "Digital Interpolation of Discrete Images," *IEEE Transactions on Computers*, Vol. C-25(2), pp. 196-202 (1976).
11. W. K. Pratt, *Digital Imaging Processing*, 2nd Edition, pp 112-117, Wiley, New York (1991).
12. S. K. Park and R. A. Schowengerdt, "Image Reconstruction by Parametric Cubic Convolution," *Computer Vision Graphics and Image Processing*, Vol. 23, pp 258-272 (1983).
13. D. H. Pritchard, "Standards and Recommended Practices," in *Television Engineering Handbook*, K. B. Benson, ed., pp. 21.10 - 21.11, McGraw-Hill, New York (1986).

7

RECONSTRUCTION

Reconstruction is the process of converting digital data into continuous analog data. It is equivalent to interpolating the signal at extremely fine increments such that the signal appears continuous. While digital interpolators are evaluated only at discrete locations, reconstruction interpolators are used continuously. That is, the analog signal can be envisioned as an extremely large number of digital points whose separation approaches zero.

In the simplest c/d/c system (See Figure 1-16, page 23), v(t) undergoes two operations. The first occurs during the sampling process:

$$v_{SAMPLE}(t) = v(t)s(t) \ . \tag{7-1}$$

The second occurs during reconstruction:

$$v_{RECON}(t) = v_{SAMPLE}(t) * h_{RECON}(t) = [v(t)s(t)] * h_{RECON}(t) \ , \tag{7-2}$$

where $h_{RECON}(t)$ is the impulse response of the reconstruction filter and $v_{RECON}(t)$ is the analog signal after reconstruction. In the frequency domain,

$$V_{RECON}(f) = V_{SAMPLE}(f) H_{RECON}(f) = [V(f) * S(f)] H_{RECON}(f) \ . \tag{7-3}$$

When satisfying the sampling theorem $v_{RECON}(t)$ will be equal to v(t). Equivalently, $V_{RECON}(f) = V(f)$. Here, S(f) and $H_{RECON}(f)$ must be unity over the frequencies that are components of V(f) and zero elsewhere.

The function $v_{RECON}(t)$ will deviate from v(t) for several reasons. If v(t) is not band-limited, aliasing occurs. This is easy to show in the frequency domain. In the base band, $V_{SAMPLE}(f)$ is corrupted by the replicated spectra and this distorts v(t). If $H_{RECON}(f)$ passes replicated frequencies, $v_{RECON}(t)$ is again distorted. When $H_{RECON}(f)$ or S(f) attenuates in-band frequency amplitudes, edges are softened. The difference between $v_{RECON}(t)$ and v(t) is discussed in Chapter 8, *Reconstructed Signal Appearance*.

In two dimensions, the displayed image is

$$i_D(x,y) = [o(x,y)s(x,y)] ** h_{RECON}(x,y) \qquad (7\text{-}4)$$

and

$$I_D(u,v) = [O(u,v) ** S(u,v)] H_{RECON}(u,v) . \qquad (7\text{-}5)$$

If $H_{RECON}(u,v)$ or $S(u,v)$ attenuates in-band frequency amplitudes, imagery appears blurry. The difference between $I_D(u,v)$ and $O(u,v)$ is often used as a measure of image quality (discussed in Chapter 11, *Image Quality Metrics*).

Electronic imaging systems may use an electronic low-pass filter, but most rely on the display medium and human visual system (HVS) response to produce a *perceived* continuous image. Display media include laser printers, halftoning, fax machines, cathode ray tubes (CRT), and flat panel displays. The display cannot be considered alone but must be coupled with the HVS response. It is the combination that makes the image appear continuous. Because the observer is usually the final interpreter of image quality, the reconstruction process is more complex than that suggested by sampling theory. Reconstruction is not used in machine vision systems where datels are manipulated by software.

The display medium paints the image onto a screen as a series of spots. By overlapping the spots, the data appear continuous. The goal is to deliberately increase the spot size to suppress the discrete nature of the data. But as the overlap increases, image detail is lost. These are competing requirements.

Many end-to-end systems are designed to be either display-limited or observer-limited. When display-limited, the disel is the largest "-el." If observer-limited, the minimum perceivable size, the HVS resel, limits resolution. In either case, the observer cannot see all the detail in the image. If too far away, he can move toward the display to see detail. But detail can only increase up to a point where the end-to-end system becomes display-limited. At this close distance, the observer can see the individual disels. The only way to see image detail is to zoom the image using an appropriate interpolation algorithm and then to display the image.

Image restoration is a process significantly different from reconstruction. It attempts to undo degradations that occur during the creation, processing, and transmission of an image. Hundreds of image restoration algorithms exist. Their effectiveness depends, in part, on the reconstruction filter characteristics. As a result, image restoration and reconstruction are treated together in the literature.

160 *SAMPLING, ALIASING, and DATA FIDELITY*

The combination provides better image (data) fidelity. This chapter covers only reconstruction filters.

The symbols used in this book are summarized in the *Symbol List* (page xiii) which appears after the *Table of Contents*.

7.1. TIME DOMAIN RECONSTRUCTION

Conceptually, a digital-to-analog converter (DAC) transforms the digital signal into a discrete analog signal (assumed to be a voltage). The reconstruction filter then converts the discrete samples into a continuous analog signal (Figure 7-1). If the digital data reside in a storage medium (e.g., hard disk, CD, floppy disk), the DAC clock rate can be any value. However, it is usually matched to the ADC clock rate to create a signal whose time-varying properties match v(t).

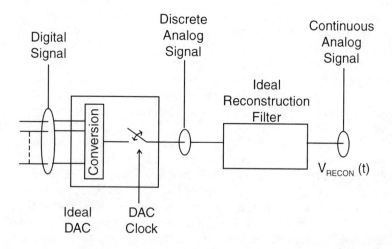

Figure 7-1. Ideal DAC with an external ideal reconstruction filter. Real DACs contain a reconstruction filter. LSB and MSB are the least- and most-significant bits.

RECONSTRUCTION 161

The ideal reconstruction filter is a low-pass filter that has unity amplitude up to the Nyquist frequency and then zero (Figure 7-2). It attenuates all the high frequency replicas created by the sampling process. In the time domain, the filter response is a sinc function. The reconstructed signal is

$$v_{RECON}(t) = \sum_{n=-\infty}^{\infty} v_{SAMPLED}(nt_s) \, \text{sinc}\left(\frac{t - nt_s}{t_s}\right). \qquad (7\text{-}6)$$

Equivalently, $v_{SAMPLE}(nt_s)$ is convolved with a sinc function. It is not possible to evaluate the equation because it requires all the original data points (the sum goes from $-\infty$ to ∞). It is necessary to truncate the sum. Equivalently, the ideal filter is not physically possible but can be approximated by many other filters such as an equi-ripple Chebyshev filter (discussed in Section 9.5.1., *Low-Pass Filter*).

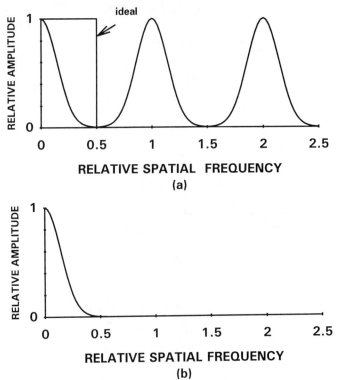

Figure 7-2. Frequency components. (a) Sampled signal and ideal filter. (b) Resultant analog signal frequency amplitude. The sampling frequency has been normalized to 1 and $f_N = 0.5$.

162 SAMPLING, ALIASING, and DATA FIDELITY

As illustrated in Figure 7-3, a typical DAC contains a zero-order reconstruction filter (sample-and-hold capability). Electronic signals are causal. No output can occur before an input. Without time delay, the output must be predicted from previous signals. For a zero-order filter, h(t) is

$$h_{RECON}(t) = 1 \quad \text{when } 0 < t < T$$
$$= 0 \quad \text{elsewhere} . \quad (7\text{-}7)$$

The Fourier transform provides

$$H_{RECON}(f) = \int_0^T e^{-j2\pi ft} dt = T \frac{\sin(\pi Tf)}{\pi Tf} e^{-j2\pi\left(\frac{T}{2}\right)f} . \quad (7\text{-}8)$$

As indicated in Figure 7-4, the output is delayed by T/2. This also illustrates the simple, back-of-the-envelope approach: the signal level is constant from one data point to the next. It makes the data look blocky.

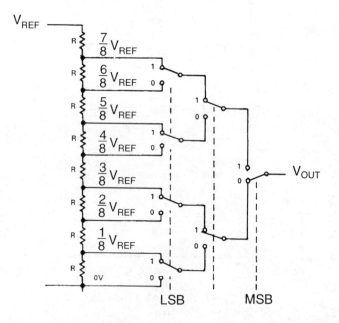

Figure 7-3. Fully decoded DAC. The digital input is 011 or 3 DN. The analog output is 3/8 V_{REF}.

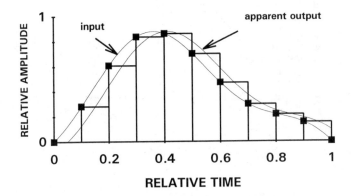

Figure 7-4. The heavy dots indicate the sampled data. Reconstructed analog signal with zero-order (sample-and-hold) filter.

A first-order filter approximates the output with straight line segments. With extrapolation, the slope of each segment is determined by the current sample and the previous sample. This creates a saw-tooth output (Figure 7-5). With delay, the slope is computed (interpolated) between two points to provide the output shown in Figure 7-6. This filter "connects the dots."

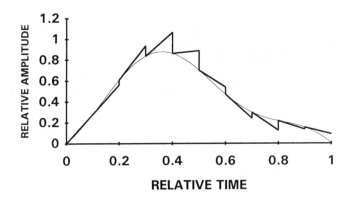

Figure 7-5. Saw-tooth output produced by a first-order filter using extrapolation.

164 SAMPLING, ALIASING, and DATA FIDELITY

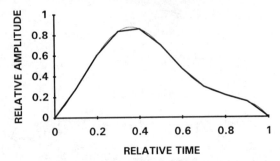

Figure 7-6. Reconstructed analog signal using a first-order reconstruction filter when delay is available.

Figures 7-7 and 7-8 illustrate the frequency response of these filters. The band-limited signal is replicated at $\pm nf_S$. Remnants of the high frequency replicas may exist with a nonideal reconstruction filter. They are spurious frequencies not in the original signal and represent a degradation of signal fidelity. Often, this artifact is erroneously labeled as "aliasing." It is not aliasing because better reconstruction filters can remove it. As the filter order increases, the analog image appears more like the original. The blockiness of these signals is caused by the reconstruction of replicated frequencies. It can be minimized with an additional low-pass reconstruction filter.

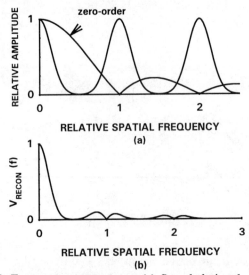

Figure 7-7. Frequency components. (a) Sampled signal and zero-order filter. (b) Resultant analog signal frequency amplitude, $V_{RECON}(f)$. The signal is band-limited and replicated at $\pm nf_S$. $f_S = 1$. The reconstructed frequency components above f_N create the blocky signal.

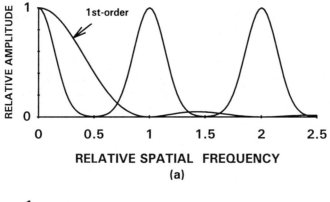

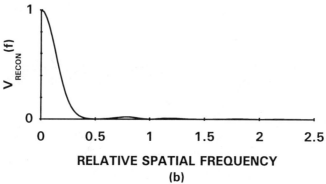

Figure 7-8. Frequency components. (a) Sampled signal and first-order filter. (b) Resultant analog signal frequency amplitude, $V_{RECON}(f)$. The signal is band-limited and replicated at $\pm nf_S$. $f_S = 1$. With minimal frequency components above f_N, $V_{RECON}(f)$ is similar to $V(f)$. Therefore $v_{RECON}(t)$ is also similar to $v(t)$.

When aliasing is present, two choices are available for reconstruction. If the filter cutoff is selected at the Nyquist frequency, then the aliased signal appears in the final signal (Figure 7-9a). The filter cutoff could be reduced (Figure 7-9b) to avoid aliasing. But this reduces the in-band signal, and the sharp cutoff results in the Gibbs phenomenon (see Section 2.1., *Fourier Series*, page 31). The Gibbs phenomenon can be moderated by a more gradual transition from the pass band to stop band. While a gradual transition reduces ringing, it also softens edges. Thus, a tradeoff is required between edge sharpness and acceptable ringing. The choice depends on the application.

Higher sampling rates simplify filter design (Figure 7-10). The effective sampling frequency can also be increased through interpolation. If the remnants

166 SAMPLING, ALIASING, and DATA FIDELITY

at nf_{S1} are reduced dramatically by digital filtering, then a very simple low-pass filter can be used for reconstruction.

Compact disks are digitally mastered by using digital interpolation to create four samples from an original sample ($m_S = 4$). The signal is then digitally low-pass filtered to remove the intermediate spectra. This simplifies the CD player's electronic design by allowing a relatively simple reconstruction filter. This is possible because the transition zone is much wider from the pass band to the replicated frequencies centered on $4f_S$. If originally band-limited, oversampling does not improve audio fidelity. It only reduces the reconstruction circuit complexity.

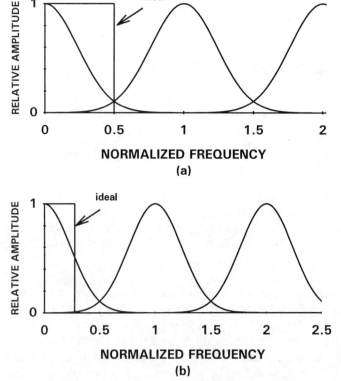

Figure 7-9. Two filter choices. (a) Cutoff at Nyquist frequency and (b) cutoff to reduce aliasing. The choice is application specific.

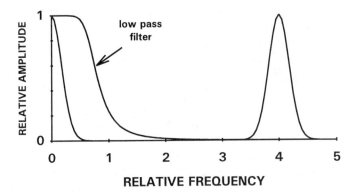

Figure 7-10. A sharp cutoff reconstruction filter is required when the highest signal frequency is near the Nyquist frequency. As the sampling frequency increases, the filter design becomes simpler. With $f_H = 0.5$, the sampling theorem requires $f_S = 1$. With $4\times$ oversampling, $f_S = 4$ (illustrated).

7.2. THE OBSERVER

The human visual system's detection capability depends on the visual angle subtended by the target size. As shown in Figure 7-11, the HVS threshold modulation (minimum perceivable modulation) is characteristically J-shaped. The HVS is most sensitive to spatial frequencies[1] that range between 3 and 5 cycles/deg at typical ambient lighting levels. The increase in threshold at low frequencies is due to the HVS inhibitory signal processing component.

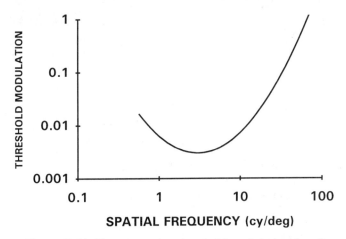

Figure 7-11. Representative threshold modulation function.

168 SAMPLING, ALIASING, and DATA FIDELITY

Some researchers have taken the inverse of the threshold modulation, normalized it, and labeled it as the HVS MTF. Although the HVS operates logarithmically, the MTF is plotted on a linear scale. As a reconstruction filter, only the high spatial frequency response needs to be modeled. An approximation that ignores inhibition is[2]:

$$MTF_{HVS}(f_{eye}) = e^{-\Gamma \frac{f_{eye}}{17.45}}, \qquad (7\text{-}9)$$

where Γ is a parameter that varies with light level (discussed in Section 9.8., *The Observer*) and is approximately unity at typical display intensities. The observer spatial frequency, f_{eye}, has units of cycles/deg and depends on the visual angle subtended by the image. Therefore, f_{eye} depends on the image size and viewing distance. Figure 7-12 illustrates MTF_{HVS} when $\Gamma = 1$.

Figure 7-13 illustrates a sweep frequency target and the apparent intensity perceived by the observer. Any image modulation approximately 60 cycles/deg or greater cannot be perceived. Therefore, a bar pattern, with a fundamental frequency of 60 cycles/deg, will be perceived as a continuous gray tone due to the spatial integration afforded by the HVS. This fact is fundamental to display medium design. When equated to observer spatial frequency, the displayed image Nyquist frequency must be greater than 60 cycles/deg for the eye to act as a good reconstruction filter. However, the observer's response is not very sharp, and therefore the in-band frequency amplitudes are reduced.

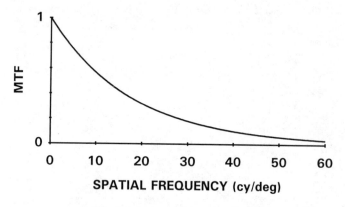

Figure 7-12. Representative $MTF_{HVS}(f_{eye})$ as a function of f_{eye}. $\Gamma = 1$.

Figure 7-14 illustrates the frequency spectrum of a square wave. When viewed at a close distance, many harmonics are within the HVS's spatial response (Figure 7-14a) and the target is perceived as a square wave. When viewed at a large distance, the harmonics are outside the HVS response (Figure 7-14b) and the target appears uniform in intensity.

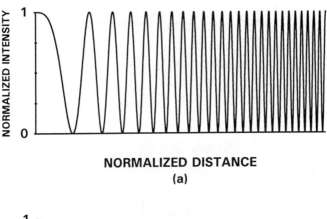

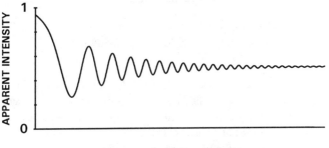

Figure 7-13. (a) Sweep frequency target and (b) the perceived intensity. When the target spatial frequency is greater than the HVS frequency response, the target is perceived as a uniform gray level (e.g., no modulation). Because the observer's frequency response is measured in units of cycles/deg, the viewing distance determines what sized target is perceivable.

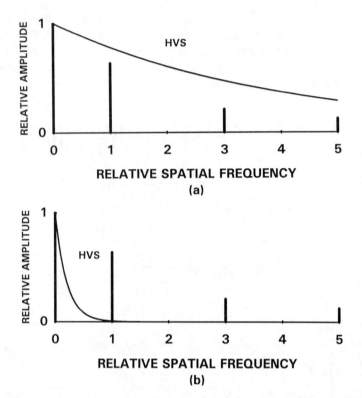

Figure 7-14. The ability to see detail depends on the viewing distance. The heavy lines are the frequency components of a square wave and the HVS frequency response is the light line. Plotted as a function of the square wave fundamental frequency. (a) At normal viewing distance, the displayed image looks like a square wave and (b) at a large distance (very high spatial frequencies presented to the eye), the square wave appears as a uniform intensity. That is, only the DC component falls within the HVS's spatial response.

7.3. RECONSTRUCTION BY THE DISPLAY MEDIUM

The human visual system does not provide a significantly sharp low-pass filter to separate the base band from the sampling frequency (i.e., first side band). Therefore, the display medium must minimize the sampling replicas.

The display medium creates an image by painting a series of (1) light spots onto a screen or film or (2) ink spots onto paper. Because of their finite size, the spot acts as low-pass filters. In fact, the goal is to increase the spot size deliberately to suppress the sampling lattice (Figure 7-15). This low-pass filter limits the amount of image sharpening possible through image enhancement or restoration.

All display media are sampling devices. They create individual spots that the HVS blends to create an apparent continuous image. Because the discrete location of the detectors has sampled the image, the display medium acts as a resampler. Any image size can be created (Figure 7-16). The only requirement is that the ratio of spot diameter to separation remains constant. This ratio is called[3] the resolution/addressability ratio or RAR.

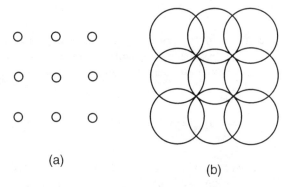

Figure 7-15. (a) With a small spot the individual datum becomes visible. Equivalently, the sampling lattice becomes visible. (b) With a large spot, the sampling lattice is suppressed.

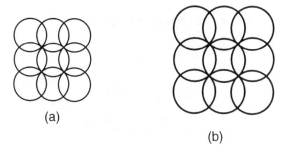

Figure 7-16. (a) Small image and (b) large image. The separation between the spots determines the overall displayed image size. As the separation increases, the spot diameter must increase in the same proportion to maintain image quality. That is, the RAR must remain constant.

172 SAMPLING, ALIASING, and DATA FIDELITY

7.3.1. RASTER SCANNED CRTs

The television camera and receiver were designed to a common video standard that specified the number of vertical lines. The signal was analog in the horizontal direction but the raster sampled the vertical direction. These were raster scanned devices. Sampling and aliasing effects were well known[4] in the early days of monochrome television.

Figure 7-17 illustrates a series of raster lines separated by a distance of d_{RASTER} mm. The vertical sampling frequency is $v_{rs} = 1/d_{RASTER}$ in units of cycles/mm. If the spot size is very large (Figure 7-18b), then the spots overlap and the raster pattern cannot be perceived. If the spot size is small (Figure 7-18a), the raster pattern becomes visible.

Spot size is a complex function of design parameters, phosphor choice, and operating conditions. It is reasonable to assume that the spot intensity profile on a CRT is Gaussian distributed. Assuming symmetry and separability,

$$i_{CRT}(x,y) \approx e^{-\frac{1}{2}\left(\frac{x^2+y^2}{\sigma_{SPOT}^2}\right)}, \qquad (7\text{-}10)$$

where σ_{SPOT} is the standard deviation of the Gaussian beam profile and x and y represent distance on the display. These parameters typically are measured in millimeters. The right hand side of Figure 7-18 illustrates the Gaussian intensity profiles.

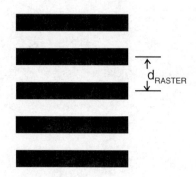

Figure 7-17. Raster scanned CRT.

RECONSTRUCTION 173

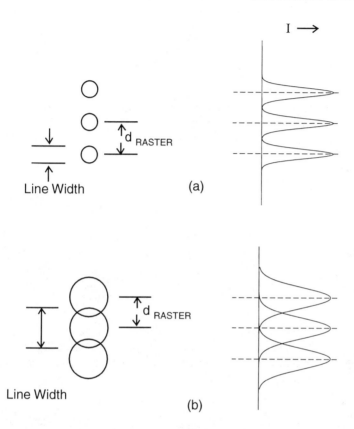

Figure 7-18. Overlapping spots act as a low-pass filter. (a) Small spots and (b) large spots. As the spot diameter increases, the raster visibility decreases. The flat field condition occurs when the lines can no longer be perceived at normal viewing distances. The left side represents the spot diameter as viewed by an observer and the right side is the intensity profile.

The intensity at any point in the vertical direction is a sum of consecutive Gaussian curves. Considering just the two nearest raster lines, the maximum intensity is the

$$I_{MAX} = 1 + 2e^{-\frac{1}{2}\left(\frac{d_{RASTER}}{\sigma_{SPOT}}\right)^2} \tag{7-11}$$

174 SAMPLING, ALIASING, and DATA FIDELITY

and the minimum intensity, which occurs between the raster lines, is

$$I_{MIN} = 2e^{-\frac{1}{2}\left(\frac{d_{RASTER}}{2\sigma_{SPOT}}\right)^2}. \tag{7-12}$$

Under nominal viewing conditions, experienced observers can no longer perceive the raster when the luminance variation (ripple) is less than 5%. This is called the flat field condition. The spot size has been standardized[5] to $\sigma_{SPOT} = d_{RASTER}/2$. The sum of closely spaced Gaussian spots is not sinusoidal. However, using the definition MTF = $(I_{MAX} - I_{MIN})/(I_{MAX} + I_{MIN})$ provides MTF = 2.3% when $\sigma_{SPOT} = d_{RASTER}/2$

The display MTF is a composite MTF that includes both the internal amplifier response and the CRT response. Implicit in the MTF is the conversion from input voltage to output display brightness. With separability, the MTF is

$$MTF_{CRT}(u_d, v_d) = e^{-2\pi^2 \sigma_{SPOT}^2 (u_d + v_d)^2}, \tag{7-13}$$

where u_d and v_d are the display spatial frequency with units of cycles/mm when measured on the display. As the spot size increases, the MTF decreases. Therefore, a tradeoff exists between image sharpness and raster pattern visibility.

Figure 7-19 illustrates a display where the Gaussian beam diameter is very small relative to d_{RASTER}. The remnants of the sampling process (all frequencies replicated about $\pm nv_{rs}$) make the raster visible.

In Figure 7-18b, the spot diameter is enlarged so that the image appears continuous. The associated MTFs are illustrated in Figure 7-20. The display MTF has suppressed the frequencies created by the sampling process but also has significantly attenuated the in-band signal. This suppression limits the amount of image sharpening possible. Some higher frequencies remain, but the HVS response can remove these.

RECONSTRUCTION 175

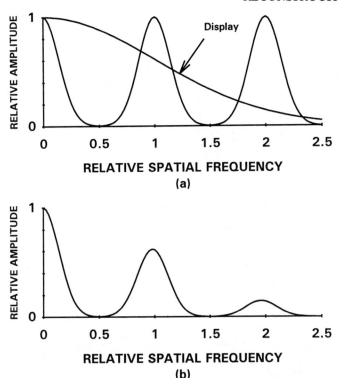

Figure 7-19. (a) Vertical response associated with small beam diameter (Figure 7-18a) and (b) resultant spectrum. The signal spectrum is band-limited to illustrate the display MTF effects. The spectrum is replicated about $\pm nv_{rs}$ where $v_{rs} = 1$.

The frequency replicas about $\pm nv_{rs}$ (Figure 7-19b) can be attenuated by the observer if he is sufficiently far enough away from the display. At close distances (Figure 7-21a) the observer can perceive remnants of the sampling process, whereas at large distances (Figure 7-21b) he cannot. By knowing the size of the target and viewing distance, the observer spatial frequency, f_{eye}, can be scaled into display spatial frequency, u_d (discussed in Section 9.1., *Frequency Domain*). The perceived signal is the signal multiplied by the display and HVS MTFs. A large viewing distance significantly attenuates the in-band amplitudes. Neither the display nor the HVS is a perfect reconstruction filter but must work together to create an image free of sampling artifacts. As illustrated in Figure 7-21c, this combination attenuates in-band frequency amplitudes.

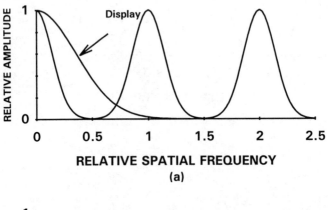

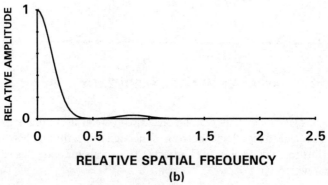

Figure 7-20. (a) Vertical response associated with large beam diameter (Figure 7-18b) and (b) resultant spectrum. The signal spectrum is band-limited to illustrate the display MTF effects. The spectrum is replicated about $\pm n v_{rS}$ where $v_{rS} = 1$.

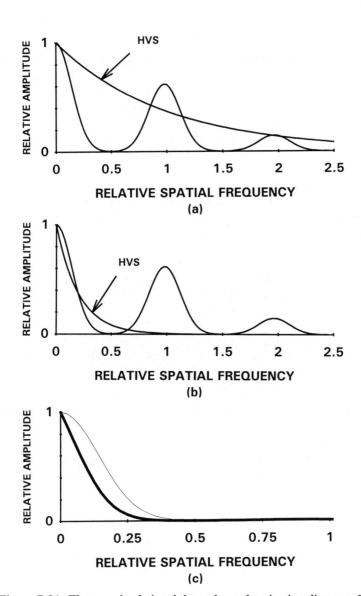

Figure 7-21. The perceived signal depends on the viewing distance. The spectrum is from Figure 7-19b. (a) At close distances, the sampling frequencies are evident. (b) At large distances sampling frequencies can be eliminated but the in-band amplitudes are significantly reduced. (c) Original (light line) and perceived signal (heavy line) at large distances.

178 SAMPLING, ALIASING, and DATA FIDELITY

7.3.2. DIGITALLY ADDRESSED CRTs

Digital displays present individual addressable spots as shown in Figure 7-22. Assuming symmetry, the sampling frequencies are $u_{ds} = v_{ds} = 1/d_{SPOT}$. As with raster scanned CRTs, the spots may be considered Gaussian (Equation 7-10, page 172). The rationale used to suppress visibility of the raster pattern also applies to a digitally addressed display. The spot diameter must be sufficiently large so that the individual spots cannot be perceived. Typically $\sigma_{SPOT} = d_{SPOT}/2$.

Raster scanned CRTs are analog devices in the horizontal direction. The digitized signal was returned to the analog domain by the camera's internal reconstruction filter. With digitally addressed CRTs, the signal is re-digitized in the horizontal direction. Therefore, Figures 7-18 through 7-21 apply also to the horizontal direction of digitally addressed CRTs.

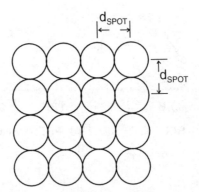

Figure 7-22. Spots on a digitally addressed CRT.

7.3.3. COLOR CRTs

A color monitor has three electron beams that scan the CRT face plate. Near the face plate is a metal screen or shadow mask that has a regular pattern of holes (Figure 7-23). The three dots create a triad or "color pixel."

The observer must be far enough away so that the individual dots of the three primaries are imperceptible. Otherwise, the individual color spots become visible and the impression of full color gamut is lost. The color dots can always be seen with appropriate magnification. A particular set of dots on a line can be considered as an infinite rectangular wave (Figure 7-24). At normal viewing distances, MTF_{HVS} restricts frequency components to values lower than the rectangular wave fundamental (Figure 7-14b, page 170). Here, the bars appear continuous. As the observer moves closer (Figure 7-14a), MTF_{HVS} starts to encompass the frequency components of the rectangular wave and the dots become obvious.

The spot diameter on all CRT-based displays is assumed to be Gaussian shaped. Although the colored dots create three sampling frequencies, one for each color, these sampling effects can be ignored. The large spot acts as a low-pass filter that attenuates the amplitude of the "color" sampling frequencies and its associated replications.

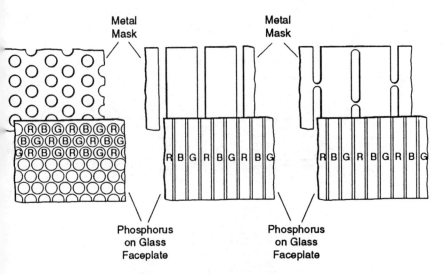

Figure 7-23. Color CRT mask patterns. (a) Dotted, (b) stripes, and (c) slotted. Most consumer television receivers use slots, and computer displays use dots. The shadow mask design varies with manufacturer.

Figure 7-24. Color dots on a line and rectangular wave representation of the green dots.

7.3.4. FLAT PANEL DISPLAYS

Flat panel displays also offer a variety of "color pixel" arrangements (Figure 7-25). The exact geometry is unimportant if the observer is far enough from the display[6,7] so that his spatial response blends the dots into an apparent continuous response. The color elements are physically separate devices and can always be seen with appropriate magnification.

Flat panel displays act as zero-order reconstruction filters. Assuming rectangular elements, the MTF is a sinc function (Figure 7-26). The HVS must attenuate the remnants of the sampling replicas to create a perceived continuous image. That is, the observer must be sufficiently far away from the display so the sampling remnants are below visibility (see Figure 7-21b, page 177).

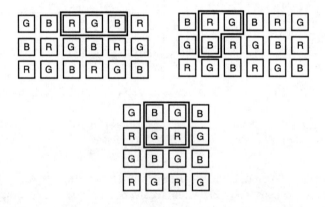

Figure 7-25. Three flat panel display geometries. The "color pixel" is outlined.

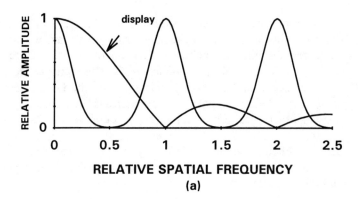

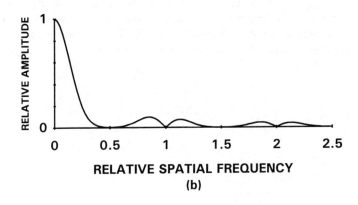

Figure 7-26. (a) Response associated with a flat panel display and (b) resultant spectrum. The input signal is band-limited to illustrate the display effects. The sampling frequency is determined by the color pixel center-to-center spacing. The spatial frequency axis has been normalized to f_s.

182 SAMPLING, ALIASING, and DATA FIDELITY

7.3.5. LASER PRINTERS

Zero-order reconstruction filters produce blocky images. Laser printers partially overcome this distortion through subscanning. The data are resampled at a high rate and the intermediate spectra are removed. The data are then reconstructed as a series of small spots (Figure 7-27). This increases the number of disels per datel, and the operation is performed by the laser printer. The number of datels within the computer memory remains the same. Figure 7-28 provides the amplitude spectrum as perceived by the observer. In effect, the image looks truly continuous. The observer must use a magnifier to see the sampling lattice. The shape of the spot MTF depends on the paper quality. With porous paper, the ink spreads radially to produce a Gaussian shaped spot. This reduces the high frequency response and edge sharpness.

Figures 7-27 and 7-28 provide a global approach to laser printer operation. Actual printers use a variety of nonlinear techniques to improve image quality and minimize the visibility of replicated spectra. These rules, which are forms of localized resampling, are nonlinear operators and can only be analyzed on a case-by-case basis.

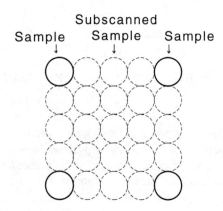

Figure 7-27. Subscanning increases the number of disels per datel.

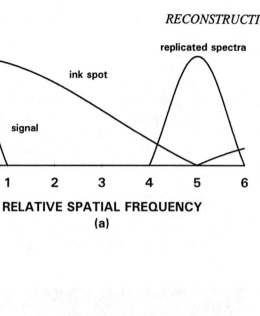
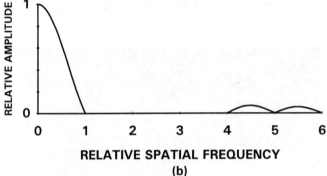
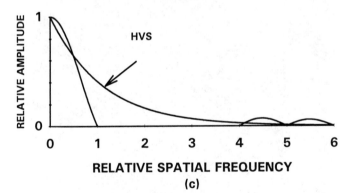

Figure 7-28. Signal amplitude spectrum modified by the ink spot. With five additional data points, $m_S = 5$. (a) A circular spot is modeled as a sinc function in one dimension. (b) The spectrum presented to the observer. (c) The observer does not significantly attenuate in-band amplitudes. f_{eye} was scaled into u_d.

184 SAMPLING, ALIASING, and DATA FIDELITY

7.3.6. HALFTONES

With printed halftone imagery, ink is either present or not present. Careful examination (with a magnifier) will reveal the individual ink dots (Figure 7-29). The ink dot sampling frequency is the inverse of the dot center-to-center spacing. By choosing an appropriate distance away from the printed page, the HVS blends the individual dots to present a perceived continuous image. Here, the ink dot sampling frequency is greater than the HVS spatial frequency limit.

The gray level, which is not present in the original (our choices are only ink or no ink), occurs because the HVS integrates over several dots. The perceived intensity is the average value of the various ink dots and remaining white space. As the dot density increases, the average value increases to give the appearance of increasing opacity.

Figure 7-29. An enlarged halftone image used in printing. When viewed at a distance of 10 ft, the individual disels are below the HVS's resolution and the tones appear continuous.

7.4. ELECTRONIC ZOOM

Zoom creates a new array that contains $m_S N \times m_S M$ datels. With one-to-one mapping onto disels, the new displayed image is $m_S \times m_S$ times larger than the original. Image quality of a zoomed image depends on the interpolation algorithm used. Datel replication is the easiest to implement and 2× replication often produces an acceptable image.[8] With more replication, the image appears blocky due to the presence of the intermediate spectra (Figure 7-30). This is sometimes called pixellation effects. When a large intensity difference exists between disel values, the HVS system perceives Mach bands[9] at the disel edges and this further accentuates the individual disels (Figure 7-31).

As a zero-order interpolator, datel replication creates a sinc function in the frequency domain. Figure 7-32 illustrates a 5× datel replication where the sinc function reduces the intermediate spectra amplitude somewhat. Figure 7-33 illustrates the 5× replication with the CRT display and HVS as reconstruction filters. The CRT and HVS spatial frequencies have been transposed to image space. With zoom, the image frequencies are reduced relative to the HVS and display frequency responses. The remnants of the intermediate frequencies create a blocky image. By physically moving farther away from the display, the HVS can remove the intermediate spectra and the image will appear continuous to the observer. This seems counterproductive. Zoom enlarges the image so that the observer can see more detail and yet he moves away to reduce blockiness. In this situation, image sharpness is limited by the observer with the display MTF having minimal effect on the image detail.

Because image frequencies are reduced compared to the HVS and display MTFs, different reconstruction filters can be evaluated. Further, the original image fidelity can be maintained by interpolating the signal, removing intermediate spectra, and then mapping the zoomed image (one-to-one) onto the display medium. An efficient interpolator has a high MTF for the in-band frequencies and reduces the amplitude of out-of-band frequencies (Figures 7-34 and 7-35). The zoomed image can be sharpened (increased MTF) whereas the original cannot due to the display medium MTF. With this process, image detail is preserved and can be perceived by the observer.

186 SAMPLING, ALIASING, and DATA FIDELITY

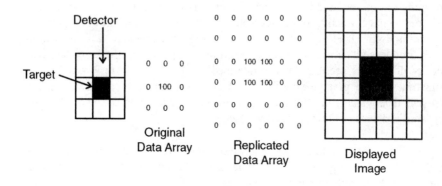

Figure 7-30. Datel replication. (Upper) Simple image and data array sizes. (Lower) Each successive image is twice the previous image. Blockiness becomes obvious when m_s is greater than two. From Reference 10.

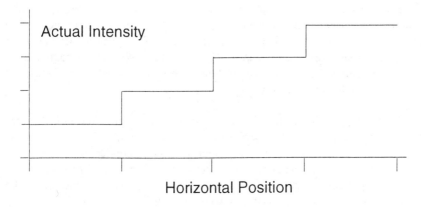

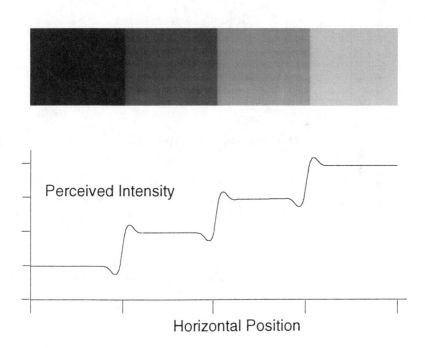

Figure 7-31. An example of the Mach band effect. A microdensitometer recording (top) of the gray scale (middle). The HVS perceives accentuated edges (bottom). From Reference 11.

188 SAMPLING, ALIASING, and DATA FIDELITY

Any combination of image processing algorithms can be used to create this process. For example, after datel replication, the image can be blurred with a Gaussian filter to reduce intermediate spectra. Then a boost filter, such as the unsharp mask,[12] is used to increase the in-band MTFs to counteract the effect of the Gaussian filter. The combination of these two can approximate an efficient interpolator. The removal of the intermediate frequencies is sometimes erroneously called anti-aliasing.

Finally, electronic zoom is different from optical zoom. With optical zoom, the focal length and field-of-view change. Because the number of samples across the field-of-view remains constant, optical zoom increases the number of samples across a target feature. This increases the electronic imaging system's resolution. With electronic zoom, the number of pixels remains constant but the number of datels increases. System resolution does not change.

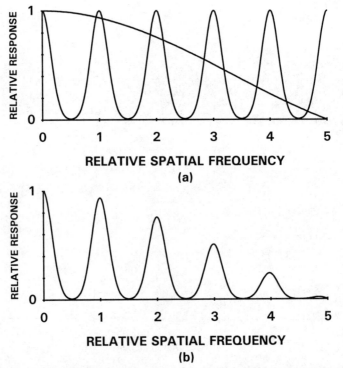

Figure 7-32. Electronic zoom ($m_S = 5$) with datel replication. (a) Sampled spectrum with intermediate spectra and sinc interpolation MTF. (b) Resultant amplitude spectrum. The signal is assumed to be band-limited. Aliasing is discussed in Section 7.5., *System Aliasing*.

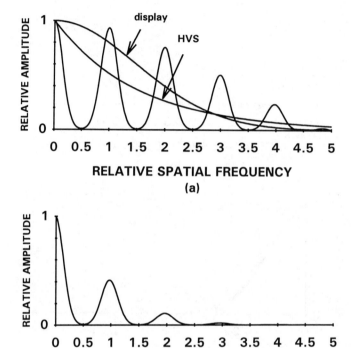

Figure 7-33. Spectrum from Figure 7-32b. (a) CRT display and HVS added and (b) perceived spectrum. The HVS and display MTFs have been scaled 5× in frequency.

190 SAMPLING, ALIASING, and DATA FIDELITY

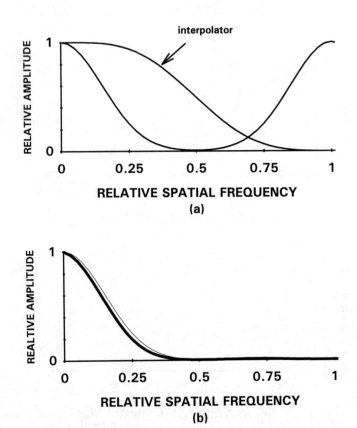

Figure 7-34. Parametric cubic convolution with $\alpha = -0.5$. (a) Individual responses and (b) resultant spectrum.

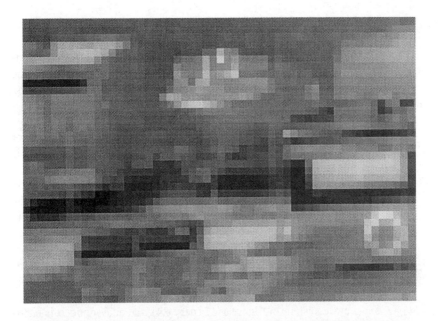

Figure 7-35. Zoomed (upper) and interpolated (lower) images. (From reference 13).

192 SAMPLING, ALIASING, and DATA FIDELITY

7.5. SYSTEM ALIASING

The previous sections showed how reconstruction filters affected band-limited signals. In addition, the frequency replications arising from electronic zoom were also illustrated with a band-limited signal. Detectors alias signals and the amount of aliasing depends on the scene content and the optical system.

Sampling creates new frequencies centered on $\pm nu_S$. Figure 7-36 illustrates only the first frequency replication (n = 1). The full detector response is illustrated in Figure 5-16 (page 115). The amplitude of the higher order replications (n > 1) are often attenuated by the optical system MTF. Figure 7-36b illustrates a system where the CRT-based spot is small. Both remnants of the sampling process and aliased signal are reconstructed. In Figure 7-36c, the remnants are eliminated but the in-band amplitudes are significantly affected with a large spot. Simultaneously, aliased components are attenuated. This leads to a tradeoff between image sharpness and aliased signal. The perceived sharpness depends on the HVS MTF.

For the c/d/c/d/c system in Figure 1-17 (page 24), the analog video is band-limited by the camera's reconstruction filter. The cutoff frequency usually is equal to a value specified by the video standard. This signal is than re-digitized by the frame grabber. Figure 6-22c (page 156) illustrated the spectrum after resampling. Note that the Nyquist frequency of the resampled data may not be the same as the detector array Nyquist frequency. The display MTF determines if remnants of the sampling process are reconstructed.

Figure 7-37 illustrates electronic zoom with the detector MTF added. For clarity, the MTF is plotted up to the detector cutoff only. Figure 7-37a shows the detector's replicated spectra and this should be compared to Figure 7-32b. The CRT and HVS modify the spectrum to create the perceived spectra (Figure 7-37b). This should be compared to the band-limited situation illustrated in Figure 7-33b. As the observer moves away from the display, the HVS further attenuates sampling remnants. This attenuates in-band frequency amplitudes.

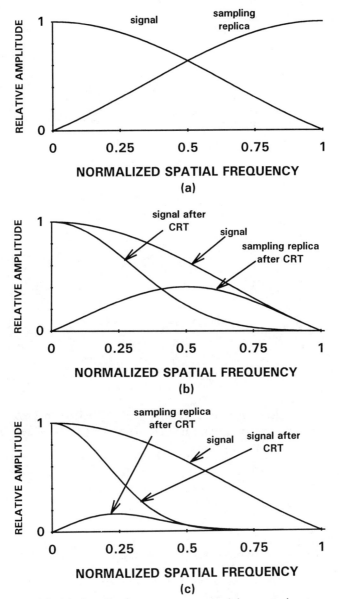

Figure 7-36. (a) Amplitude spectrum created by a staring array with 100% fill-factor. (b) When the CRT spot size is small, frequencies above u_N are reproduced. (c) Large spots significantly attenuate in-band (up to u_N) signal amplitudes.

194 SAMPLING, ALIASING, and DATA FIDELITY

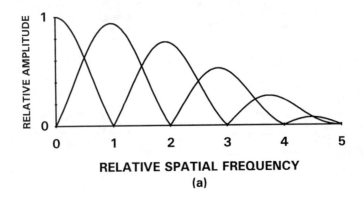

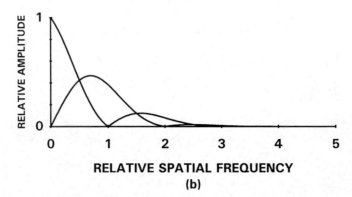

Figure 7-37. Electronic zoom with $m_S = 5$. (a) Spectrum before the display and (b) perceived spectrum. The detector MTF is plotted up to the normalized cutoff ($u_{iD} = 1$).

7.6. REFERENCES

1. B. O. Hultgren, "Subjective Quality Factor Revisited," in *Human Vision and Electronic Imaging: Models, Methods and Applications*, B. E. Rogowitz and J. P. Allebach, eds., SPIE Proceedings Vol. 1249, pp. 12-22 (1990).
2. G. H. Kornfeld and W. R. Lawson, "Visual Perception Model," *Journal of the Optical Society of America*, Vol. 61(6), pp. 811-820 (1971).
3. G. C. Holst, *CCD Arrays, Cameras, and Displays*, pp. 189-194, JCD Publishing, Winter Park, FL (1996).
4. P. Mertz and F. Gray, "A Theory of Scanning and its Relation to the Characteristics of the Transmitted Signal in Telephotography and Television," *Bell Systems Technical Journal*, Vol. XIII, pp. 464-515 (1934).
5. L. M. Biberman, "Image Quality," in *Perception of Displayed Information*, L. M. Biberman, ed., pp. 13-18, Plenum Press, New York (1973).

6. T. Hase, "Two Dimensional Modulation Transfer Function of Luminance on Display Images With a Mosaic Pixel Structure," *IEEE Transactions on Consumer Electronics*, Vol. 40(2), pp. 83-91 (1994).
7. R. J. Beaton, "Linear Systems Metrics of Image Quality for Flat-Panel Displays," in *Image Processing, Analysis, Measurements, and Quality*, SPIE Proceedings Vol. 901, pp. 44-151 (1988).
8. C. J. Woodruff and G. N. Newsam, "Displaying Undersampled Imagery," *Optical Engineering*, Vol. 23(2), pp. 579-585 (1994).
9. A. Fiorentini, "Mach Band Phenomena," in *Handbook of Sensory Physiology*, D. Jameson and L. M. Hurvich, eds., Vol. VII-4, pp. 188-201, Springer-Verlag, Berlin (1972).
10. A. R. Weeks, Jr., *Fundamentals of Electronic Image Processing*, page 28, SPIE Optical Engineering Press, Bellingham, WA (1996).
11. A. R. Weeks, Jr., *Fundamentals of Electronic Image Processing*, page 23, SPIE Optical Engineering Press, Bellingham, WA (1996).
12. A. R. Weeks, Jr., *Fundamentals of Electronic Image Processing*, pp. 139-144, SPIE Optical Engineering Press, Bellingham, WA (1996).
13. R.C. Hardie, S. Cain, K. J. Barnard, J. Bognar, E. Armstrong, and E. A. Watson, "High Resolution Image Reconstruction From a Sequence of Rotated and Translated Infrared Images," in *Infrared Imaging Systems: Design, Analysis, Modeling, and Testing VIII*, G. C. Holst, ed., SPIE Proceedings Vol. 3063, pp. 113-124 (1997).

8
RECONSTRUCTED SIGNAL APPEARANCE

The three-bar and four-bar patterns have become popular test targets. They are characterized by their fundamental frequency only. The expansion of a square wave into a Fourier series clearly shows that it consists of an infinite number of sinusoidal frequencies. Although the square wave fundamental frequency may be oversampled, the higher harmonics may not. During digitization, the higher order frequencies will be aliased down to lower frequencies and the square wave will change its appearance. There will be intensity variations from bar-to-bar and the bar width will not remain constant. The appearance of the reconstructed signal depends on the reconstruction filter used. With nonideal reconstruction, periodic targets may appear to have beat frequencies.

Using the Nyquist frequency as the limit of system performance severely restricts system applications. Nyquist frequency applies to sinusoidal signals. Many studies illustrate sampling effects with sinusoids or square waves (characterized by its fundamental frequency - a sinusoid). The real world contains aperiodic targets and signals. As such, samples per pulse and samples on target may be more meaningful.

An edge location can only be determined within one sample time. Phasing effects become obvious when f_o/f_N is not an integer. Many computer simulations and hand-drawn diagrams use integers so that the detector locations are either in-phase or out-of-phase with respect to the target. Therefore beat frequencies and aliasing effects are not often seen with quickly drawn diagrams. A different reconstruction filter (higher order, display, or HVS) will modify the reconstructed signal appearance. Which graphical representation to use depends on the specific system being evaluated.

Just examining the sampler or MTF in isolation does not provide complete information. With imagery, the perceived signal depends on the optical MTF, the spatial integration afforded by the detector, sampling effects, display MTF, and the observer's distance from the display. Blurring produced by the optics and that caused by sampling and reconstruction are fundamentally distinct with respect to their origin. However, it may be difficult to differentiate their presence in the displayed image.

RECONSTRUCTED SIGNAL APPEARANCE 197

The symbols used in this book are summarized in the *Symbol List* (page xiii) which appears after the *Table of Contents*.

8.1. PHASING EFFECTS

When the Nyquist frequency is an integer value of the sinusoid period, phasing effects are more obvious (e.g., $f_N = f_o$, $f_N = 2f_o$, etc.). When $f_N = f_o$ there are exactly two samples per period. Figure 8-1 illustrates an "out-of-phase" condition and Figure 8-2 illustrates the in-phase condition where the signal is sampled at its maximum and minimum by a flash ADC. The output of an ADC is a series of digital numbers that reside in a computer memory. To make these data viewable, a zero-order reconstruction filter was used. The difference between Figures 8-1 and 8-2 is that the sampler phase has changed by $\pi/2$ with respect to the signal. Whether the maximum or minimum is captured is a "lucky shot."

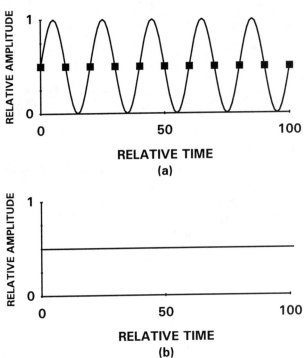

Figure 8-1. A sinusoid sampled at Nyquist frequency. (a) Sampled at the average value and (b) apparent output after a zero-order reconstruction filter. This is an "out-of-phase" relationship. The heavy dots indicate sample values.

198 SAMPLING, ALIASING, and DATA FIDELITY

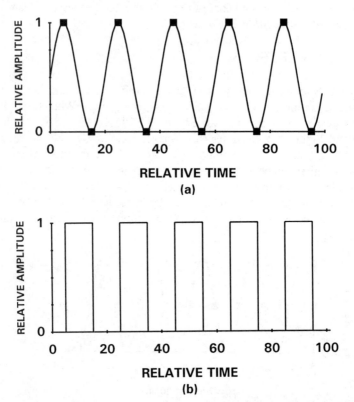

Figure 8-2. A sinusoid sampled at Nyquist frequency. By changing the sampling phase by $\pi/2$ the signal output is maximized. (a) Sampled at the maximum and minimum values and (b) apparent output after a zero-order reconstruction filter. This is the in-phase relationship. An ideal reconstruction filter will return a sinusoidal signal. The heavy dots indicate sample values.

Phasing effects do not modify the sampling theorem. Except for the singular case where the out-of-phase sampling provided no output (Figure 8-1), the signal can always be reconstructed as a sinusoid with the ideal low-pass reconstruction filter.

Figures 8-3 and 8-4 illustrate a sinusoid sampled three times per cycle. Both zero-order and first-order reconstruction filter responses are provided for two different phases. The two different reconstruction filters create what appear to be different signals. In both cases, these inefficient filters allow remnants of the sampling process to remain in the reconstructed amplitude spectrum. The ideal reconstruction filter will create a sinusoid that is a replica of the input. Deviation from the ideal filter response introduces the signal distortions seen in these figures.

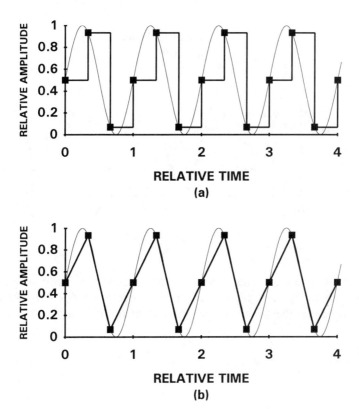

Figure 8-3. Three samples per cycle. Apparent output of (a) zero-order and (b) first-order reconstruction filters. The heavy dots indicate sample values.

200 SAMPLING, ALIASING, and DATA FIDELITY

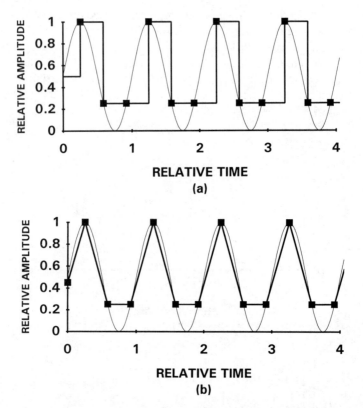

Figure 8-4. Phasing effects change the signal appearance. Three samples per cycle. Apparent output of (a) zero-order and (b) first-order reconstruction filters. The heavy dots indicate sample values.

Figure 8-5a illustrates a square wave sampled at 2.22 times the fundamental. The phasing effects are clearly visible. In Figure 8-5b, a zero-order reconstruction filter converts the digital data into an analog signal. The pulse width and apparent location vary from pulse to pulse. Never does ambiguity extend past one sample. The heavy dots indicate the sample values and represent the data available to an image processing algorithm.

Variations in edge location affect all image processing algorithms. The variations in pulse width can only be seen on a display because the digital data reside in computer memory. Note that output pulse widths are equally spaced. They just do not match up with the input signal.

Figures 8-1 through 8-5 typify systems that use either zero- or first-order reconstruction filters. These filters are found in a variety of instruments such as digital oscilloscopes. Higher sampling frequencies minimize the phasing effects.

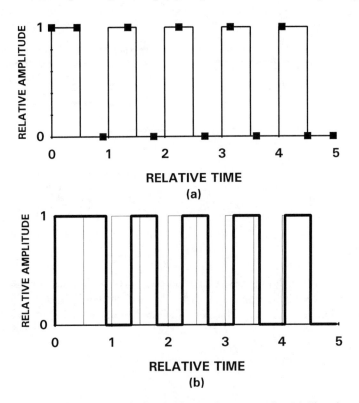

Figure 8-5. Square sampled at 2.22 samples per cycle. (a) Signal and (b) analog output (heavy line) after a zero-order reconstruction filter. The heavy dots indicate sample values.

8.2. EDGE AMBIGUITY

As shown in Figure 8-5, sampling introduces ambiguity in target edge location. This is also true with imaging systems but with the added feature that the detector integrates over a finite area. Figure 8-6 illustrates an edge and detector array output after a zero-order reconstruction filter. This figure typifies an imaging system output where the optical blur diameter is very small compared to the detector size. Many CCD-based imaging systems with low f-number lens systems fall into this category. The detectors are contiguous and

202 SAMPLING, ALIASING, and DATA FIDELITY

the analog output is created by a zero-order reconstruction filter. As the relative phase changes between the edge location and detector array, the detector outputs change. The perceived signal depends on the display MTF and the observer's distance from the display.

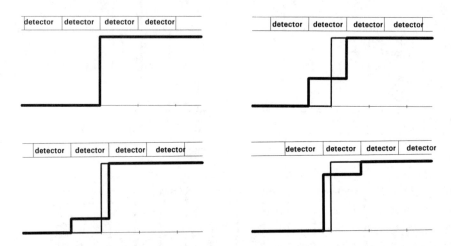

Figure 8-6. Phasing effects with a detector array. The detectors are contiguous and the analog signal (heavy line) is created by a zero-order reconstruction filter. The display MTF and HVS spatial integration softens the edges of the perceived image. The light line is the edge impinging on the detector array.

Figure 8-7 illustrates an edge when the optical blur is equal to the detector size. This typically applies to CCD imaging systems with high f-number lens systems and most infrared imaging systems. As the optical blur diameter increases, the system changes from being detector- to optics-limited. Equivalently, resolution changes from the pixel to the optical resel. As the blur diameter increases, the edge width increases.

The edge location with respect to the detector location is random. Figure 8-8 illustrates the range of responses using a first-order reconstruction filter as a function of linear fill-factor. That is, the reconstructed signal will fall somewhere inside the shaded area. The total shaded area may be considered a measure of the reconstructed image's fidelity.[1] In this figure, the blur diameter is small compared to a detector width. As the blur diameter increases, the shaded area also increases.

RECONSTRUCTED SIGNAL APPEARANCE 203

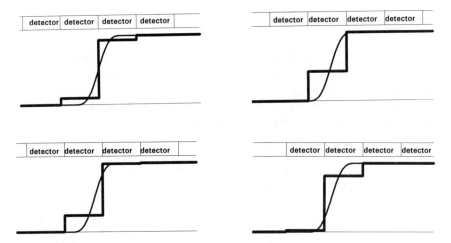

Figure 8-7. Phasing effects with a detector array when the optical blur diameter is equal to the detector size. The detectors are contiguous and the analog signal (heavy line) is created by a zero-order reconstruction filter. The light line is the edge impinging on the detector array. It is softened by optical MTF.

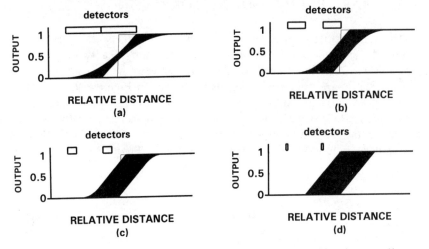

Figure 8-8. The shaded area is the ambiguity introduced by the sampling process. The edge is at the center, and the detector outputs (after a first-order reconstruction filter) vary according to phase. Range of responses for different linear fill-factors: (a) 100%, (b) 50%, (c) 25%, and (d) 0%. (d) is identical to a flash ADC. The linear fill-factor is the detector width divided by the center-to-center spacing (detector pitch).

8.3. TARGET WIDTH AMBIGUITY

The variation in edge location leads to the question: What is the smallest signal (flaw, blob or dot) that can be detected? Equivalently, What is the minimum number of samples required across the target? For the flash ADC, the signal width must be greater than one sample time to ensure detection. Otherwise, a lucky shot is needed to detect it.

When viewing a point source, diffraction-limited optics produces a diffraction pattern that consists of a bull's-eye pattern with concentric rings. The center is the Airy disk and its diameter is

$$d_{AIRY} = 2.44 \frac{\lambda fl}{D_o} = 2.44 \lambda F .\qquad (8\text{-}1)$$

When d_{AIRY} is small compared to the detector size, the detector response controls the signal appearance. When d_{AIRY} is large, the optical blur significantly affects the signal appearance.

8.3.1. SMALL BLUR-TO-DETECTOR RATIO

Detectors spatially integrate the signal, and targets can move both horizontally and vertically. Therefore, the minimum is a two-dimensional consideration. Detection depends on the phase between the signal and detector array and the image processing algorithm selected. In Figure 8-9a, a circular flaw produces an image that just fills one detector. In Figure 8-9b, the image is shifted by one-half detector width both horizontally and vertically. If fully illuminated, the maximum detector output is 255 with an 8-bit ADC. Because the area of a circle is $\pi D^2/4$. the digitized outputs of Figure 8-9a create a data array:

$$\begin{matrix} 0 & 0 & 0 & 0 \\ 0 & 200 & 0 & 0 \\ 0 & 0 & 0 & 0 \\ 0 & 0 & 0 & 0 \end{matrix} .\qquad (8\text{-}2)$$

RECONSTRUCTED SIGNAL APPEARANCE 205

The outputs for Figure 8-9b are placed into a data array:

$$\begin{matrix} 0 & 0 & 0 & 0 \\ 0 & 50 & 50 & 0 \\ 0 & 50 & 50 & 0 \\ 0 & 0 & 0 & 0 \end{matrix} \quad (8\text{-}3)$$

Any graphical representation of the data assumes a specific reconstruction filter. The displayed image depends on the display medium. For flat panel displays, Figure 8-9a would appear as a 1 × 1 square disel. With CRTs, Figure 8-9a appears as a Gaussian shaped pulse and Figure 8-9b will appear as a flattened Gaussian pulse. The flaw can be detected by an image processing algorithm, if the threshold is set sufficiently low. If the threshold is set at 25, the flaw will appear as one datel in Figure 8-9a and four datels in Figure 8-9b after processing. If the threshold is set at 100, the flaw will appear as one datel in Figure 8-9a and not be present in Figure 8-9b.

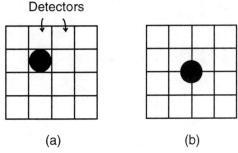

(a)   (b)

Figure 8-9. A circular flaw that is equal to the detector size. (a) Flaw centered on one detector and (b) flaw centered on the junction of four detectors.

Figure 8-10 illustrates a blob whose area covers three detectors. The digitized outputs of Figure 8-10a are placed into a data array:

$$\begin{matrix} 0 & 0 & 0 & 0 & 0 & 0 \\ 0 & 135 & 248 & 135 & 0 & 0 \\ 0 & 248 & 255 & 248 & 0 & 0 \\ 0 & 135 & 248 & 248 & 0 & 0 \\ 0 & 0 & 0 & 0 & 0 & 0 \\ 0 & 0 & 0 & 0 & 0 & 0 \end{matrix} \quad (8\text{-}4)$$

206 SAMPLING, ALIASING, and DATA FIDELITY

The data array for Figure 8-10b is

$$\begin{matrix} 0 & 0 & 0 & 0 & 0 & 0 \\ 0 & 2 & 97 & 97 & 2 & 0 \\ 0 & 97 & 255 & 255 & 97 & 0 \\ 0 & 97 & 255 & 255 & 97 & 0 \\ 0 & 2 & 97 & 97 & 2 & 0 \\ 0 & 0 & 0 & 0 & 0 & 0 \end{matrix}$$ (8-5)

Depending on the threshold selected, the flaw can be one to four datels wide after processing. For finite fill-factor arrays, the blob diameter must be greater then the diagonal distance between two detectors (Figure 8-11a) to ensure some detection. For one detector to be fully illuminated, the diagonal must increase as shown in Figure 8-11b. The minimum blob size depends[2] on the detector pitch and size, optical blur diameter, and the image processing algorithm.

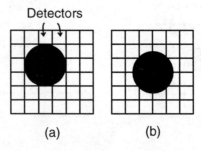

Figure 8-10. A flaw that is equal to three detector widths. Flaw centered on (a) one detector and (b) junction of four detectors.

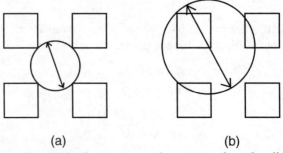

Figure 8-11. The blob diameter must be greater than the diagonal to ensure covering at least part of one detector. (a) Detector partially illuminated and (b) fully illuminated.

As the shape of the object changes, the values in the data array also change. Figure 8-12 illustrates a bar at two orientations. Depending on the phase, a narrow bar may cover one or two detectors in any direction. Image processing algorithms affect the number of datels that represent the bar. Objects at angles with respect to the detector array have a stair-step appearance after reconstruction. This affects measurement accuracy as the object size decreases. The minimum size necessary is application specific.

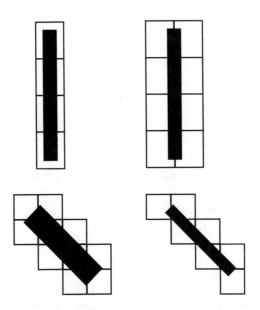

Figure 8-12. A narrow bar may cover one or two detectors in any direction. The light line indicates the detectors that will produce an output. There is always ambiguity in edge location.

8.3.2. LARGE BLUR-TO-DETECTOR RATIO

Detection of small sources, such as stars, requires a large blur diameter compared to the detector size. Figure 8-13 illustrates the phasing effects when the blur diameter is equal to four contiguous detector widths. Although phasing introduces ambiguity, a large number of detectors across the blur diameter permits accurate reconstruction. This also spreads the image intensity across many detectors so that the signal-to-noise is proportionally reduced. With most systems, the blur diameter approaches d_{AIRY}. Then the detector size and pitch is selected to ensure adequately sampling. That is, small-sized detectors are used to create a large blur-to-detector ratio.

208 SAMPLING, ALIASING, and DATA FIDELITY

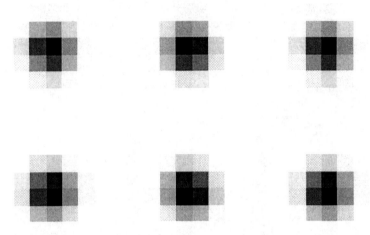

Figure 8-13. Phasing effects when the blur diameter is equal to four detector widths. (By courtesy of Office National d'Études et de Recherches Aérospatiales)

8.4. BEAT FREQUENCIES

Because the zero-order filter is often used for back-of-the-envelope diagrams, it is worthwhile to see its effects. Figure 8-14a illustrates a sinusoid sampled 10 times per period. After the digital-to-analog converter (neglecting possible quantization effects), the signal is passed though a zero-order filter. The resultant wave form is shown in Figure 8-14b. Figure 8-14c provides the amplitude spectrum after the zero-order filter. The input frequency is replicated at $nf_s \pm f$. With a sampling frequency of $f_s = 1$, the input frequency of $f_o = 0.1$ is replicated at 0.9, 1.1, 1.9, 2.1, $\cdots$.

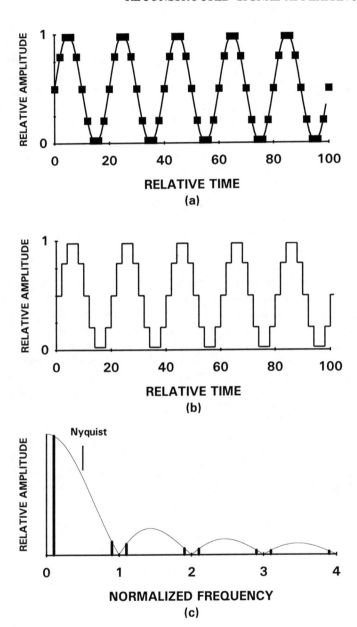

Figure 8-14. Sinusoid sampled at 10 times per period. (a) Sinusoid and sample locations (heavy dots), (b) zero-order output, and (c) amplitude spectrum after the zero-order filter.

210 SAMPLING, ALIASING, and DATA FIDELITY

In Figure 8-15, the sampling frequency has been decreased to 3.333 times per period. In Figure 8-16, the sampling frequency is 2.222 per period. The beat frequency noticeable in Figure 8-16b is due to the interaction of f_o with $f_S - f_o$. Its frequency is $f_{BEAT} = (f_S - f_o) - f_o = f_S - 2f_o = f_N - f_o$. The beat frequency period lasts for N_{CYCLE} input frequency cycles (Figure 8-17):

$$N_{CYCLE} = \frac{f_o}{2(f_N - f_o)} . \quad (8\text{-}6)$$

As f_o/f_N approaches one, the beat frequency becomes obvious. Compare the relative spacing between f_o and $f_S - f_o$ in Figures 8-14c, 8-15c, and 8-16c. As the distance decreases, the beat frequency is more obvious (compare Figures 8-14b, 8-15b, and 8-16b).

Careful examination of the frequency spectra (Figures 8-14c, 8-15c, and 8-16c) reveal that the distortion is due to the beat frequency only and not to aliasing or any other artifact. An ideal low-pass reconstruction filter (unity amplitude up to f_N and zero after that) will eliminate the higher order frequencies. With the ideal reconstruction filter, only the fundamental frequency will remain and the reconstructed signal will be a replica of the original signal appearing as a sinusoid. Figures 8-14 through 8-16 violate a fundamental requirement of the sampling theorem: they do not use an ideal reconstruction filter.

The beat frequencies are a result of the sampling process and the use of a nonideal reconstruction filter. This is unfortunate. The simple zero-order concept is extremely easy to use for back-of-the-envelope analysis. It may not represent actual system design.

A beat frequency becomes obvious when (a) the input frequency is near the Nyquist frequency and (b) an inefficient reconstruction filter is used. The same phenomenon occurs with displays. Figure 8-18a illustrates the replicated frequencies at $v_{rS} - v_o$. With small spots (Figure 8-18b) the replicated frequency is reconstructed and beat frequencies will be perceived, whereas with large spots (Figure 8-18c) they will not.

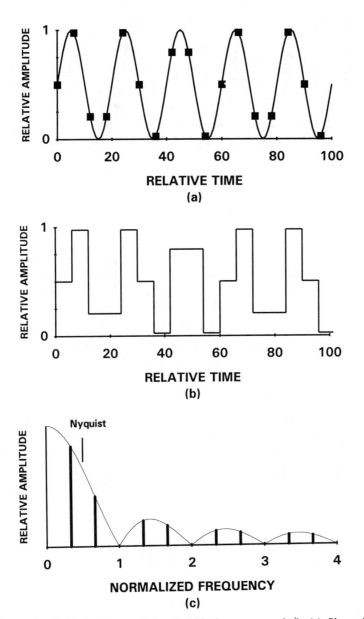

Figure 8-15. Sinusoid sampled at 3.333 times per period. (a) Sinusoid and sample locations, (b) zero-order output, and (c) amplitude spectrum after the zero-order filter.

212 SAMPLING, ALIASING, and DATA FIDELITY

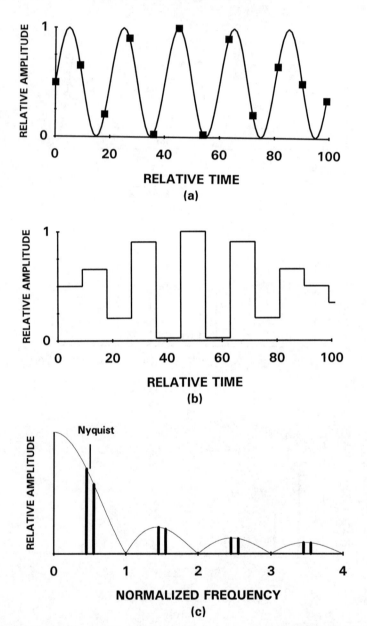

Figure 8-16. Sinusoid sampled at 2.22 times per period. (a) Sinusoid and sample locations, (b) zero-order output, and (c) amplitude spectrum after the zero-order filter.

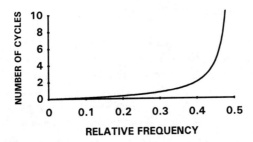

Figure 8-17. Number of input frequency cycles required to see one complete beat frequency cycle.

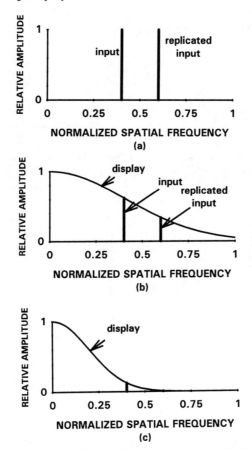

Figure 8-18. (a) Replicated frequencies at $nv_{rs} \pm v_o$. (b) Small spot CRTs will reproduce $v_{rs} - v_o$. (c) Large spot CRTs attenuate the signal at $v_{rs} - v_o$ and no beat frequency is seen.

214 SAMPLING, ALIASING, and DATA FIDELITY

8.5. BAR TARGET APPEARANCE

The square wave (bar pattern) is the most popular test target and is usually described by its fundamental frequency only. The expansion of a square wave into a Fourier series clearly shows that it consists of an infinite number of sinusoidal frequencies. Although the square wave fundamental frequency may be oversampled, some higher harmonics will not. During digitization, the higher order frequencies will be aliased down to lower frequencies and the square wave will change its appearance. Aliasing will create intensity variations from bar-to-bar. The displayed output bar width depends on the relationship between the detector location and the bar location.

To understand sampling effects,[3] a one-dimensional square wave of infinite extent was evaluated analytically. The fundamental frequency is $u_o/u_N = 0.952$. Figure 8-19 illustrates the aliasing of $3u_o$, $5u_o$, $7u_o$, $9u_o$, and $11u_o$ into the base band.

For this study, the optical MTF was assumed to be unity over the spatial frequencies of interest. The bar frequency amplitudes are modified by the detector MTF as illustrated in Figure 8-20. Figure 8-21 shows the frequency spectrum after a zero-order reconstruction filter. If an ideal reconstruction filter was added, the spectrum would only contain frequencies up to u_N.

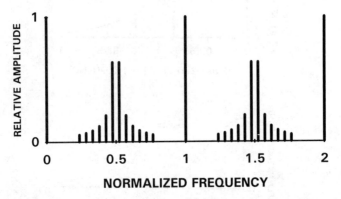

Figure 8-19. Aliased bar pattern frequency spectra when $u/u_N = 0.952$. Only frequencies up to the 11th harmonic are shown: $uf_o/u_N = 2.856$, $u_o/u_N = 4.76$, $7u_o/u_N = 6.664$, $9u_o/u_N = 8.568$, and $11u_o/u_N = 10.472$. The frequencies are replicated about $\pm nu_S$, where $u_S = 1$. The DC value is the average intensity of the entire bar pattern.

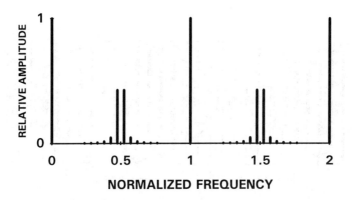

Figure 8-20. Aliased square wave frequency spectrum modified by the detector MTF.

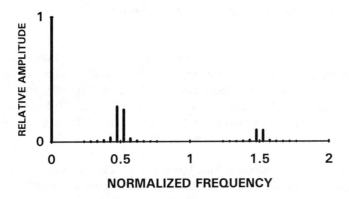

Figure 8-21. Frequency spectrum after the zero-order reconstruction filter. The dominant frequencies at u_o and $u_S - u_o$ create a beat frequency in the displayed bar pattern.

If $u_o/u_N = 0.952$, the beat frequency is equal to 9.9 cycles of the input frequency (Figure 8-22). Here, the target must contain at least 10 cycles to see the entire beat pattern. Aliased components distort the pulse. In these figures, a zero-order reconstruction filter was used. With the ideal reconstruction filter, the beat frequencies disappear and the pulse appears as a distorted sinusoid. Many electronic imaging systems do not have the ideal filter so beat patterns are possible. They may be minimized by the display MTF or HVS.

216 SAMPLING, ALIASING, and DATA FIDELITY

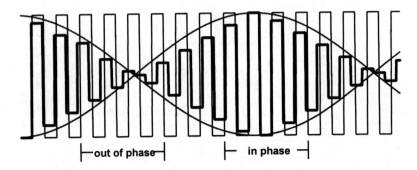

Figure 8-22. Beat frequency produced by an ideal staring system when $u_o/u_N = 0.952$. $N_{CYCLE} = 9.9$ and $d_H/d_{CCH} = 1$. The light line is the input and the heavy line is the detector output. The beat frequency envelope is shown. Figure 8-16 illustrates the frequency spectrum.

Standard characterization targets consist of three or four bars. The image of the bar pattern would have to be moved $\pm\frac{1}{2}$ d (detector extent) to change the output from a maximum value (in-phase) to a minimum value (out-of-phase). This can be proven by selecting just four bars in Figure 8-22. Depending on the phase, a four-bar pattern may either be replicated or nearly disappear. With only a few bars, the beat pattern will not be seen.

When u_o/u_N is less than about 0.6 (Figure 8-23), the beat frequency is not obvious. Now the output nearly replicates the input but slight variation in pulse width and amplitude occurs. In the region where u_o/u_N is between approximately 0.6 and 0.9, adjacent bar amplitudes are always less than the input amplitude (Figure 8-24).

When u_o/u_N is less than about 0.6, a four-bar pattern will always be seen (select any four adjoining bars in Figure 8-23). When $u_o/u_N < 0.6$, phasing effects are minimal and a phase adjustment of $\pm\frac{1}{2}d$ in image space will not affect an observer's ability to resolve a four-bar target. In the region where u_o/u_N is between approximately 0.6 and 0.9, four-bar targets will never *look* correct (Figure 8-24). One or two bars may be either wider or of lower intensity than the others.

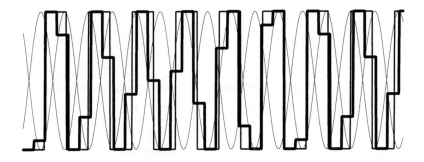

Figure 8-23. Ideal staring system output when $u_o/u_N = 0.522$. $N_{CYCLE} = 0.54$ and $d_H/d_{CCH} = 1$. The light line is the input and the heavy line is the detector output. The output nearly replicates the input when $u_o/u_N < 0.6$.

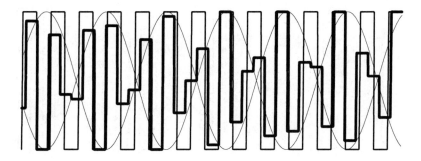

Figure 8-24. Ideal staring system output when $u_o/u_N = 0.811$. $N_{CYCLE} = 2.14$. The light line is the input and the heavy line is the detector output. The output never *looks* quite right when u_o/u_N is between 0.6 and 0.9.

Input frequencies of $u_o = u_N/k$, where k is an integer, are faithfully reproduced (i.e., no beat frequencies). When $k = 1$, as the target moves from in-phase (Figure 8-25) to out-of-phase, the output will vary from a maximum to zero. Selection of u_N/k targets avoids the beat frequency problem but significantly limits the number of spatial frequencies selected for testing.[4]

Signals whose frequencies are above Nyquist frequency will be aliased down to lower frequencies. This would be evident if an infinitely long periodic target

218 SAMPLING, ALIASING, and DATA FIDELITY

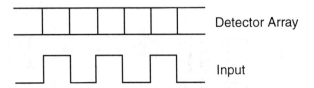

Figure 8-25. In-phase relationship where $u_o = u_N$. Careful alignment is required to ensure maximum output is obtained.

Signals whose frequencies are above Nyquist frequency will be aliased down to lower frequencies. This would be evident if an infinitely long periodic target was viewed. However, when u_o is less than about 1.15 u_N, it is possible to select a phase such that four adjoining bars *appear* to be faithfully reproduced (Figure 8-26). These targets can be *resolved* although the underlying fundamental frequency has been aliased to a lower frequency.

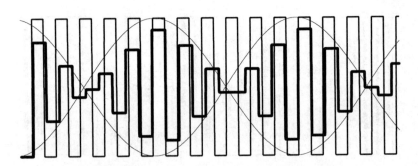

Figure 8-26. Ideal staring system output when $u_o/u_N = 1.094$. $N_{CYCLE} = 5.8$. The light line is the input and the heavy line is the detector output. By selecting the appropriate phase, the output *appears* to replicate an input four-bar pattern.

Figures 8-22 through 8-25 represent the detector output when d_H/d_{CCH} is one. As the d_H/d_{CCH} decreases, the detector MTF increases and its effect on the square wave harmonics diminishes. It is the finite width detector that emphasizes the beat frequencies. In the limit that the detector is very small, the beat frequency disappears when viewing square waves. This, then, approximates the output of a frame grabber (see Figure 8-5, page 201).

RECONSTRUCTED SIGNAL APPEARANCE 219

The sweep frequency bar target can illustrate all of these effects. As illustrated in Figure 8-27, the bar pattern frequency increases from left to right. For signals below $f_N/2$, the output pulse looks like a square wave. Between $f_N/2$ and f_N, the output is irregular. Changing the phase will change the output appearance. Above f_N, the frequencies are aliased to lower frequencies. An envelope about the output approximately follows a sinc² function. The detector spatial integration provides one sinc, and the zero-order reconstruction filter provides the second. Deviations from the sinc² function are caused by phasing effects.

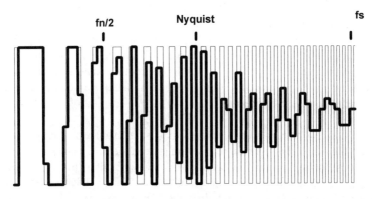

Figure 8-27. Ideal staring array output when viewing a sweep frequency bar pattern. The light line is the bar pattern and the heavy line is the output after a zero-order reconstruction filter.

The contrast transfer function (CTF) is the system response to square wave targets. For linear-shift-invariant systems, the CTF and MTF are related by simple equations.[5] The advantage of using square wave targets to calculate the MTF is their simplicity. Anyone can make them, whereas spatial sinusoids are difficult to fabricate.[6] The conversion from CTF to MTF assumes that the targets have infinite extent. Boreman and Yang[7] have illustrated the relationship between the MTF and CTF for the popular three- and four-bar targets. However, sampled data systems corrupt the bar target appearance and thereby make data interpretation difficult. They indicate that the conversion is acceptable for frequencies less than $u_N/2$. This is consistent with Figure 8-23 where it was demonstrated that the square wave is adequately reproduced when $u_o/u_N < 0.6$.

8.6. INDEPENDENT PIXELS ON TARGET

The required number of pixels on a target depends on the task. With machine vision systems there must be a sufficient number of datels for the computer algorithm to identify the target within some accuracy. The computed image size depends on both phasing effects and the threshold set in the algorithm. The threshold may be set high to provide a high signal-to-noise ratio or to avoid false positive identifications.

For many military applications an observer evaluates the image. Recognition and identification require the observer to discern target detail. Based on experience, the observer can "fill in" areas that are missing, partially obscured, or distorted. That is, the observer is tolerant of aliasing when it comes to target detection and recognition. These tasks do not require imagery that is a faithful reproduction of the scene. They only require an adequate number of pixels on the target to perform the task.

The concept of pixels on the target may be appropriate when using staring arrays. It assumes that the imaging system is detector-limited and that the optical MTF does not significantly affect the imagery. Although "pixels per target" is common parlance, the correct question is How many independent "-els" are required? The number of *independent* pixels is defined as

$$N_{TARGET} = \frac{target\ width}{range} \frac{1}{PAS}, \quad (8\text{-}7)$$

where PAS the pixel-angular-subtense and it must not be confused with the detector-angular-subtense. The minimum number of (independent) pixels per target depends on the task. As the number increases, detail increases and also the probability of recognition and identification.[8]

Figure 8-28a illustrates the silhouette of a Soviet Bear bomber. In Figures 8-28b to 8-28e the number of pixels on the target increases. These images contain only eight gray levels to emphasize sampling effects. The system MTF is considered unity over the spatial frequencies of interest. Simple pixel averaging was used to create these figures. More sophisticated image reconstruction algorithms will provide more pleasing and less blocky images. Nevertheless, this simplified approach illustrates sampling effects. Optical blur, electronic MTFs, and the reconstruction process will smooth out these images.

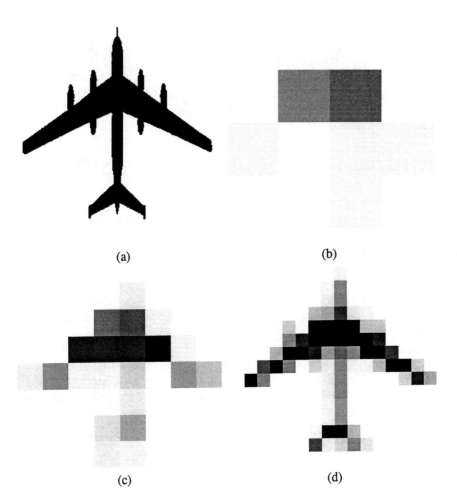

Figure 8-28. Soviet Bear bomber. (a) Silhouette. (b) Image of bomber with four contiguous pixels across the wing span. (c) Eight contiguous pixels across the wing span. (d) 16 contiguous pixels across the wing span. (continued next page)

222 SAMPLING, ALIASING, and DATA FIDELITY

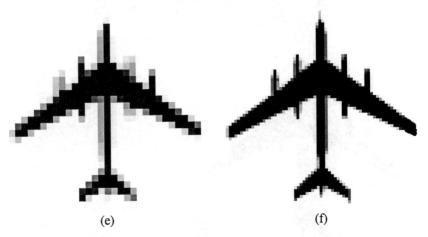

(e) (f)

Figure 8-28 (continued). Soviet Bear bomber. (e) 32 contiguous pixels across the wing span. (f) 64 contiguous pixels across the wing span. Eight levels of gray are used.

Although Figure 8-28c (eight pixels on target) does not look like a bomber, that fact is easily inferred from other information. For example, trucks and ships do not fly. Radar data would provide the range. The airplane size can be estimated from the relationship between the image size and the system field-of-view. Knowledge of the plane size and shape makes identification possible. It is the auxiliary information that is often used to identify objects - not just the object outline. Any mathematical approach considered cannot include these complex learned parameters. Thus, pixels on target are a simplification of the human interpretation process.

All figures illustrate phasing effects. For example, in Figure 8-28e, the fuselage is not of uniform intensity. It appears darker on one side than the other. The engines appear as either one pixel wide or two pixels wide. When two pixels wide, they are of reduced intensity. Although 16 pixels across the wing span (Figure 8-28d) are sufficient to identify the object as an aircraft, 32 pixels are required to discern the engines (Figure 8-28e).

Although the PAS is used in Equation 8-7, the limiting "-el" should be used. It may be created by the optical system, detector, sampling, the display, or the HVS. By using the limiting "-el," the entire system is considered rather than just one component. Independent "-els" are sometimes called uncorrelated "-els." For example, oversampling increases the number of pixels per target but does not change the number of *independent* pixels as specified by Equation 8-7.

8.7. CHARACTER RECOGNITION

The legibility of alphanumeric characters on computer monitors is related to the resolution/addressability ratio (RAR). It is related[9] to the CRT spot size and therefore the CRT MTF. Figure 8-29 illustrates three letters and the resultant intensity traces. The RAR must be near one so that a reasonable contrast ratio exists between on and off disels. With reasonable contrast, the inner detail of the character is seen and the character is legible. Similarly, with a reasonable contrast ratio, adjacent letters will appear as separate letters. With characters, the alternating disel pattern is called one stroke separated by a space. Readable characters are typically formed in a block of disels ranging from 5 × 7 to 9 × 16. With one-to-one mapping, this represents 5 × 7 to 9 × 16 pixels per character.

Most document readers employ staring arrays. Optical character readers (OCRs) generally require 200 pixels per inch (called dots per inch) on the document. Font size is specified in points where one point is approximately 1/72" (0.0139"). A 10-point font uses 0.139 inches from the top of a capital "T" to the bottom of a lower case "y." A single letter is about 3/4 of the font size so that with 10-point font, a single letter is about 0.10". With 200 dpi (0.005" per pixel), there are 20 pixels per letter. Generally, machine vision systems require higher resolution than the human observer. Therefore, 20 pixels per character is consistent with the size necessary for human readability.

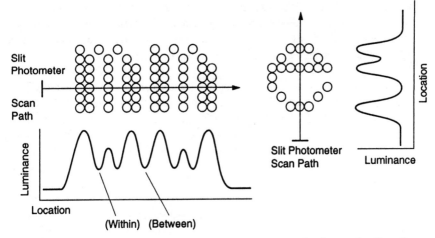

Figure 8-29. The RAR must be near one so that the inner details of characters are visible and the double width appears solid. Similarly with RAR near one, two adjacent characters appear separate. Each dot represents a disel.

224 SAMPLING, ALIASING, and DATA FIDELITY

8.8. ASYMMETRIC SAMPLING

Scanning arrays tend to have asymmetric sampling. Staring arrays may or may not have symmetrical sampling. It is generally accepted that machine vision system algorithms are easier to implement when MTFs and sampling lattices are equal in the horizontal and vertical directions. With asymmetric sampling, interpolation must be used to create a geometrically correct image (see Section 6.1., *Examples*, page 132). The observer is more tolerant of the sampling lattice differences.

Park et al.[10] analyzed the Landsat multispectral scanner system (MSS). The MSS system is a line scanner that is undersampled both in the along and across track directions. In the across track direction (scan direction), PAS = 0.76·DAS and in the along track direction, PAS = 1.08·DAS. From a mathematical point of view, this anisotropic undersampling should provide obvious image asymmetry. This 1.42:1 mismatch in sampling is less obvious when the optical blur is added to the imagery (Figure 8-30). The target is in the upper left. The upper right image is blurred due to optical diffraction and the continuous convolution of the square detector DAS. The objects in the lower left and right are more blurred images of this target because they are formed by convolving with a 1.42:1 rectangle whose size is matched to the MSS average system MTF. Two orthogonal orientations are shown. The images are blurred and do exhibit some asymmetry, but the observer is very tolerant of degraded imagery. Our visual experience allows us to "fill in" the missing parts and still identify the object as a pentagon. Additional subsystem MTFs will further degrade the system MTF and the asymmetry will be even less noticeable. Sample scene effects make this asymmetry difficult to measure.

Obert et al. reported[11] several computer simulation results in which the DAS and sampling grid were varied in both the horizontal and vertical directions. When the number of samples per DAS was reduced to less than two, the probability of correct response was reduced. However, the significant finding was that as the vertical resolution was reduced, the probability of correct response was reduced. *This specificity of dimension suggests that vertical resolution and vertical sampling rate may be slightly more important....than horizontal resolution and horizontal sampling. The results must be interpreted cautiously, however, because the task involved targets that are either square (front view) or elongated in the horizontal direction (right front, side) which would require relatively greater vertical than horizontal resolution.* They went on to say, *If the aspect ratio is considered in system design, increasing the vertical rather than the horizontal resolution through detector aspect configuration [e.g., a rectangular detector] may yield a slightly greater degree*

RECONSTRUCTED SIGNAL APPEARANCE 225

of performance improvement FOR THE TARGET VEHICLES OF THE TYPE USED IN THIS STUDY. The capitalized words (not in the original paper) are added to emphasize a potential limitation of their results.

The results suggest that it is the direction of the target's critical dimension (e.g., the smallest dimension) where the highest resolution is required. Intuitively this is reasonable. This suggests that the system application must be known before the system resolution (detector size) and sampling lattice is selected.

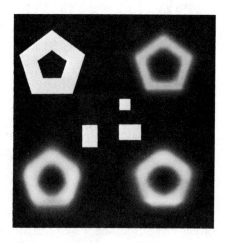

Figure 8-30. Simulated LANDSAT multispectral imagery. The asymmetric sampling of 1.42:1 is not considered visually objectionable. The rectangles represent the sampling lattice size (From Reference 10).

8.9. DYNAMIC SAMPLING

With microscan, the scene is static and the effective detector locations change which increases the sampling rate (see Section 5.3.4., *Microscan*, page 112). A similar effect occurs when the scene is moving (e.g., changing phase). The imagery will change as shown in Figure 8-31. If these frames are presented in real time (i.e., 30 frames/sec), the HVS temporally integrates the images to provide a perceived higher quality image. Barbe and Campana[12] reported the results of a CCD staring camera operating at 60 frames/sec. If target motion was greater than 1/8 of a detector width per frame, image quality improved. This is equivalent to increasing the sampling frequency from $1/d_{CC}$ to $8/d_{CC}$. For slower movements (higher effective sampling rates), no further improvement was perceived. With very fast movement, the CCD integration time blurred the image.

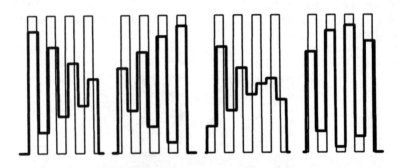

Figure 8-31. Different outputs created by a moving four-bar target. $u_o/u_N = 1.06$. If presented in real time, each bar would appear equally visible.

They also viewed the three-bar target. Although the Nyquist frequency was 16.7 cycles/mm, they could resolve targets at 50 cycles/mm. Note that this value is above the detector cutoff, u_{iD}. But the detector has response above u_{iD} (see Figure 5-5, page 105) and defining cutoff as u_{iD} artificially suggests that the detector cannot respond to higher frequencies. Webb[13] reported a similar test in which the detectability of a four-bar pattern improved. He demonstrated that it was possible to perceive a target whose fundamental frequency was 1.5 times the Nyquist frequency.

The ability to perceive higher spatial frequency targets depends on the system MTF. Webb tested an infrared system where the system MTF was significantly affected by the optical MTF, whereas Barbe and Campana were not limited by the optical MTF. As a result, the improvement seen by Webb was not as great as that seen by Barbe and Campana. Optical prefiltering reduces the MTF and the ability to perceive the higher frequency targets. Similarly, the reconstruction filter also acts as a low-pass filter. Even if the detector can reproduce these higher frequencies, they may be well above the reconstruction filter cutoff and therefore eliminated.

8.10. REFERENCES

1. S. K. Park and R. A. Schowengerdt, "Image Sampling, Reconstruction, and the Effect of Sample-scene Phasing," *Applied Optics*, 21(17), pp. 3142-3151 (1982).
2. G. C. Holst, *Electro-Optical Imaging System Performance*, pp. 48-54, JCD Publishing, Winter Park, FL (1995).
3. G. C. Holst, "Sampling, Aliasing, and Target Appearance," *Infrared Physics & Technology*, Vol. 37, pp. 627-634 (1996).
4. G. C. Holst, *Testing and Evaluation Infrared Imaging Systems*, pp. 238-246, JCD Publishing, Winter Park, FL (1993).
5. J. W. Coltman, "The Specification of Imaging Properties by Response to a Sine Wave Input," *Journal of the Optical Society of America*, Vol. 44(6), pp. 468-471 (1954).
6. Photographic sinusoids are available from Sine Patterns, 236 Henderson Drive, Penfield, NY 14526.
7. G. D. Boreman and S. Yang, "Modulation Transfer Function Measurement Using Three-and Four-bar Targets," *Applied Optics*, Vol. 34, pp. 8050-8052 (1995).
8. G. C. Holst, *Electro-Optical Imaging System Performance*, pp. 254-259 and 412-440, JCD Publishing, Winter Park, FL (1995).
9. G. C. Holst, *CCD Arrays, Cameras, and Displays*, pp. 189-194, JCD Publishing, Winter Park, FL (1996).
10. S. K. Park, R. Schowengerdt, and M. A. Kaczynski, "Modulation-transfer-function Analysis for Sampled Image Systems," *Applied Optics*, 23(15), pp. 2572-2582, (1984).
11. L. Obert, J. D'Agostino, B. O'Kane, and C. Nguyen, "An Experimental Study of the Effect of Vertical Resolution on FLIR Performance," in *Proceedings of the IRIS Specialty Group on Passive Sensors*, Vol. 1, pp. 235-251, Infrared Information Analysis Center, Ann Arbor, Mich (1990).
12. D. F. Barbe and S. B. Campana, "Imaging Arrays Using the Charge-Coupled Concept," in *Image Pickup and Display, Volume 3*, B. Kazan, ed., pp. 245-253, Academic Press (1977).
13. C. M. Webb, "MRTD, How Far Can We Stretch It?" in *Infrared Imaging Systems: Design, Analysis, Modeling, and Testing V*, G. C. Holst, ed., SPIE Proceedings Vol. 2224, pp. 297-307 (1994).

9

SYSTEM ANALYSIS

System analysis is performed in the frequency domain. As such, it does not usually include the scene nor the displayed image. Therefore, it is not directly used to determine target appearance. Rather, target appearance is inferred from the number of samples across a target feature (see Chapter 8, *Reconstructed Signal Appearance*, page 196). Of course, the displayed image can be evaluated through a simulation. With this process, the object is Fourier transformed and then multiplied by the pre-sampling MTFs. Aliasing is added and the remaining MTFs are multiplied. The inverse transform provides the simulated displayed image.

This chapter discusses aliasing as necessary to illustrate certain points. Reduced MTF only affects scenes that have frequency components at those spatial frequencies. Scenes can always be selected where a low system MTF has little effect on performing a task such as target recognition or identification. In addition, scenes can be selected to emphasize aliasing. Unfortunately, these tend to be test patterns and this complicates data analysis.

Throughout this text specific subsystem MTFs were included to illustrate certain points. Many examples used ideal components. The ideal filter is, of course, unrealizable (see Section 3.4., *Response of Idealized Circuits*, page 77). This chapter discusses real components. It also includes the MTFs previously described for completeness.

The basic c/d/c electronic imaging system consists of optics, a detector array, an electronic subsystem, and a display:

$$MTF_{SYSTEM} = MTF_{OPTICS} MTF_{DETECTOR} MTF_{ELECTRONICS} MTF_{DISPLAY} \; . \quad (9\text{-}1)$$

The observer processes this information to create a perceived MTF:

$$MTF_{PERCEIVED} = MTF_{SYSTEM} MTF_{HVS} \; . \quad (9\text{-}2)$$

MTF theory provides guidance for system design. When coupled with noise, the MTF specifies system performance for those systems used for general imagery. By using these equations, it is assumed that the object (or target) consists of all frequencies over the frequency range of interest. The highest frequency that can be faithfully reproduced is limited by the optical cutoff. The detector, electronics, display, or observer may reduce the perceived frequency content. Although MTF theory is used to describe system performance, the analyst must always be aware that the displayed image depends on the object and that aliasing is usually present. Scene aliasing and sampling effects are further discussed in Chapter 11, *Image Quality Metrics*.

Historically, the calculated system MTF did not include the display nor the observer. The primary reason was that the user supplied his own display. This is still true today. With analog devices, this is a reasonable approach. However, image reconstruction, the display, and the HVS response determine whether the sampling process is perceivable. Thus, system level MTF analysis should include these important MTFs. An MTF graph that does not specify the Nyquist frequency is incomplete. If significant spectra exist above Nyquist frequency, then the image may appear blocky or the sampling lattice may be visible.

Only the optics and detector array are sensitive to the scene spectral and spatial content. The remaining subsystem MTF components only modify the electrical signals. These remaining MTFs apply to all imaging systems independent of the system spectral response. By modifying the electrical signals created by the detector, these MTFs modify the displayed image.

This chapter considers six major MTFs: optics, detector, electronic filtering, reconstruction filters, the display, and the observer. Numerous[1,2] other MTFs may be added as necessary such as motion, detector diffusion effects, charge transfer efficiency, and TDI velocity mismatch.

The symbols used in this book are summarized in the *Symbol List* (page xiii) which appears after the *Table of Contents*.

9.1. FREQUENCY DOMAIN

There are four different locations where spatial frequency domain analysis is appropriate. They are object space (before the camera optics), image space (after the camera optics), display, and observer (at the eye) spatial frequencies. Simple equations relate the spatial frequencies in all these domains.

230 SAMPLING, ALIASING, and DATA FIDELITY

Figure 9-1 illustrates the spatial frequency associated with a bar target. Bar patterns are the most common test targets and are characterized by their fundamental spatial frequency. Using the small angle approximation, the angle subtended by one cycle (one bar and one space) is d_o/R_1, where d_o is the spatial extent of one cycle and R_1 is the distance from the imaging system entrance aperture to the target. When using a collimator to project the targets at apparently long ranges, the collimator focal length replaces R_1. Targets placed in the collimator's focal plane can be described in object space. The horizontal object-space spatial frequency, u_{ob}, is the inverse of the horizontal target angular subtense and is usually expressed in cycles/mrad:

$$u_{ob} = \frac{1}{1000}\left(\frac{R_1}{d_o}\right) \frac{cycles}{mrad} \ . \quad (9\text{-}3)$$

The object space domain is used by the military for describing system performance

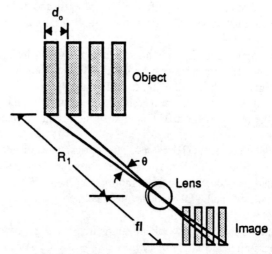

Figure 9-1. Correspondence of spatial frequencies in object and image space. Although the MTF is defined for sinusoidal signals, the bar target is the most popular test target (shown here). Bar targets are characterized by their fundamental frequency.

Optical designers typically quote spatial frequencies in image space to specify the resolving capability of lens systems. Photographic cameras and CCD cameras are typically specified with spatial frequencies in image space which is the object-space spatial frequency divided by the system focal length:

$$u_i = \frac{u_{ob}}{fl} \quad \frac{line\text{-}pairs}{mm} \quad or \quad \frac{cycles}{mm}, \qquad (9\text{-}4)$$

where u_i is the inverse of one cycle in the focal plane of the lens system. Although used interchangeably, line-pairs suggest square wave targets and cycles suggest sinusoidal targets. To maintain dimensionality, if u_{ob} is measured in cycles/mrad then the focal length must be measured in meters to obtain cycles/mm.

Analog electronic filters are one-dimensional and modify a serial data stream. The electrical frequency is related to the FOV and the time it takes to read out one line. For analog filters immediately following the detector array (read out time is $t_{H\text{-}LINE}$),

$$f_e = \frac{HFOV \cdot fl}{t_{H\text{-}LINE}} u_i \quad Hz \;. \qquad (9\text{-}5)$$

For staring arrays with N_H horizontal detectors,

$$f_e = \frac{(N_H - 1)d_{CCH} + d_H}{t_{H\text{-}LINE}} u_i \quad Hz \;. \qquad (9\text{-}6)$$

While digital filters operate on a data array, the effective sampling rate is related to the timing of the read out clock:

$$f_{eS} = \frac{1}{T_{CLOCK}} \quad Hz \;. \qquad (9\text{-}7)$$

232 SAMPLING, ALIASING, and DATA FIDELITY

After the digital-to-analog converter, the serial stream data rate and therefore the filter design is linked to the video standard. The active line time, $t_{VIDEO-LINE}$, (Table 9-1) and array size provide the link between image space and video frequencies:

$$f_v = \frac{HFOV \cdot fl}{t_{VIDEO-LINE}} u_i \quad Hz \quad . \tag{9-8}$$

For staring arrays,

$$f_v = \frac{(N_H - 1)d_{CCH} + d_H}{t_{VIDEO-LINE}} u_i \quad Hz \quad . \tag{9-9}$$

Table 9-1
STANDARD VIDEO TIMING

FORMAT	TOTAL LINE TIME (μs)	MINIMUM ACTIVE LINE TIME (μs) $t_{VIDEO-LINE}$
EIA 170	63.492	52.092
NTSC	63.555	52.456
PAL	64.0	51.7
SECAM	64.0	51.7

The field-of-view and monitor size link image spatial frequency to the horizontal and vertical display spatial frequencies:

$$u_d = \frac{HFOV \cdot fl}{W_{MONITOR}} u_i \quad \frac{cycles}{mm} \tag{9-10}$$

and

$$v_d = \frac{VFOV \cdot fl}{H_{MONITOR}} v_i \quad \frac{cycles}{mm} \quad . \tag{9-11}$$

Both the monitor width, $W_{MONITOR}$, and height, $H_{MONITOR}$, are usually measured in millimeters. When the monitor aspect ratio is the same as the FOV ratio (the usual case),

$$\frac{HFOV}{W_{MONITOR}} = \frac{VFOV}{H_{MONITOR}} . \quad (9\text{-}12)$$

The spatial frequency presented to the observer depends on the observer's viewing distance and the image size on the display (Figure 9-2). Observer response is assumed to be rotationally symmetric so that the horizontal and vertical responses are the same. The usual units are cycles/deg:

$$u_{eye} = \frac{1}{2 \tan^{-1}\left(\frac{1}{2}\frac{1}{Du_d}\right)} \quad \frac{cycles}{deg} . \quad (9\text{-}13)$$

Here, the arc tangent is expressed in degrees. Because image detail is important, the small angle approximation provides

$$u_{eye} = 0.01745\, D\, u_d \quad \frac{cycles}{deg} . \quad (9\text{-}14)$$

Table 9-2 summarizes the various spatial frequencies.

Table 9-2
TRANSFORM PAIR UNITS

SUBSYSTEM RESPONSE	VARIABLES	TRANSFORMED VARIABLES
Optics, detector	x denotes distance (mm) y denotes distance (mm)	u_i expressed as cycles/mm v_i expressed as cycles/mm
Optics, detector	θ_x denotes angle (mrad) θ_y denotes angle (mrad)	u_{ob} expressed as cycles/mrad v_{ob} expressed as cycles/mrad
Analog electronics	t denotes time (sec)	f_e expressed as cycles/sec (Hz)
Video electronics	t denotes time (sec)	f_v expressed as cycles/sec (Hz)
Display	x denotes distance (mm) y denotes distance (mm)	u_d expressed as cycles/mm v_d expressed as cycles/mm
Observer	θ_x denotes angle (deg) θ_y denotes angle (deg)	u_{eye} expressed as cycles/deg v_{eye} expressed as cycles/deg

234 SAMPLING, ALIASING, and DATA FIDELITY

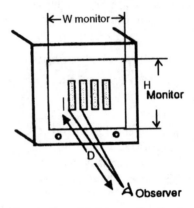

Figure 9-2. Correspondence of spatial frequencies for the display and for the eye.

9.2. OPTICS MTF

Optical systems consist of several lenses or mirror elements with varying focal lengths and indices of refraction. Multiple elements are used to minimize lens aberrations. While individual element MTFs are used for design and fabrication, these individual MTFs cannot be cascaded to obtain the optical system MTF. For modeling purposes, the optical system is treated as a single lens with the same effective focal length and aberrations as the lens system.

Optical spatial frequency is two-dimensional with the frequency ranging from $-\infty$ to $+\infty$. The diffraction-limited MTF for a circular aperture was first introduced in Section 3.5., *Superposition Applied to Optical Systems*, page 78. For an aberration free and radially symmetric optical system, MTF_{OPTICS} is the same in the horizontal and vertical directions. In the horizontal direction,

$$MTF_{OPTICS}(u_i) = \frac{2}{\pi} \left[\cos^{-1}\left(\frac{u_i}{u_{iC}}\right) - \frac{u_i}{u_{iC}} \sqrt{1 - \left(\frac{u_i}{u_{iC}}\right)^2} \right] \quad \text{when } u_i < u_{iC} \quad (9\text{-}15)$$

$$= 0 \quad \text{elsewhere}.$$

The optical cutoff is $u_{iC} = D_O/(\lambda \, fl) = 1/(F\lambda)$. D_O is the aperture diameter and fl is the effective focal length. The f-number, F, is equal to fl/D_O.

SYSTEM ANALYSIS 235

Figure 9-3 illustrates MTF_{OPTICS} as a function of u_i/u_{iC}. This equation is only valid for noncoherent monochromatic light and the cutoff frequency is dependent on the wavelength. The extension to polychromatic light is lens specific. Most lens systems are color corrected (achromatized) and therefore there is no simple way to apply this simple formula to predict the MTF. As an approximation to the polychromatic MTF, the average wavelength, is used to calculate the cutoff frequency:

$$\lambda_{ave} \approx \frac{\lambda_{MAX} + \lambda_{MIN}}{2}. \tag{9-16}$$

For example, if the system spectral response ranges from 0.4 to 0.7 μm, then $\lambda_{AVE} = 0.55$ μm.

For most electronic imaging systems sensitive in the visible spectral region, the optical system may be considered near diffraction limited and in focus. Approximations for aberrated and defocused optical systems may be found in Reference 3. The MTFs for optical systems with rectangular apertures[4] or telescopes with a central obscuration (Cassegrainian-type optics) are also available.[5]

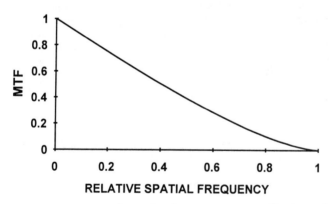

Figure 9-3. MTF_{OPTICS} for a circular aperture normalized to u_i/u_{iC}.

236 *SAMPLING, ALIASING, and DATA FIDELITY*

9.3. DETECTORS

The detector MTF was introduced in Section 5.3., *Detectors*, page 102, to illustrate the spatial integration afforded by a detector. The two-dimensional spatial response of a rectangular detector (Figure 9-4) is

$$MTF_{DETECTOR}(u_i, v_i) = |sinc(d_H u_i)||sinc(d_V v_i)| . \quad (9\text{-}17)$$

Although the detector MTF is valid for all spatial frequencies from $-\infty$ to $+\infty$, it is typically plotted up to the first zero (called the detector cutoff) which occurs at $u_{iD} = 1/d_H$. Because d_H may be different than d_V, the horizontal and vertical MTF can be different. Recall that a phase shift occurs when $u_i > d_H$ and $v_i > d_V$.

Note that arrays are often specified by the pixel size. With 100% fill-factor arrays, the pixel size is equal to the detector size. With finite fill-factor arrays, the photosensitive detector dimensions are less than the pixel dimensions. A microlens or an anti-alias filter may optically modify the effective size (see Section 5.3.8., *Optical Prefiltering*, page 122).

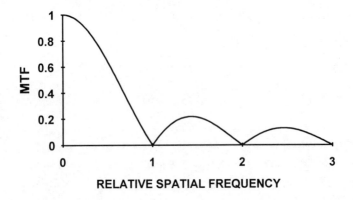

Figure 9-4. Detector MTF as a function of normalized spatial frequency $d_H u_i$. The MTF is usually plotted up to the first zero ($d_H u_i = 1$). For staring arrays, the array Nyquist frequency must be added: $u_{iN} = 1/(2d_{CCH})$.

SYSTEM ANALYSIS 237

9.4. SAMPLE-SCENE MTF

Because of phasing effects, sampled data systems do not have[6-10] a unique MTF. A wide range of MTF values is possible for any given spatial frequency. The MTF depends on the phase relationship of the scene with the sampling lattice. Superposition does not hold and any MTF derived for the sampler cannot, in principle, be used to predict results for general scenery.

To account for this variation, a sample-scene MTF may be included. Then the detector spatial frequency response is represented by $MTF_{DETECTOR}MTF_{PHASE}$. This composite MTF provides an average performance response. MTF_{PHASE} is[11]

$$MTF_{PHASE}(u_i) \approx \cos\left(\frac{u_i}{u_{iN}}\theta\right), \quad (9\text{-}18)$$

where θ is the phase angle between the target and the sampling lattice. For example, when $u_i = u_{iN}$, the MTF is a maximum when $\theta = 0$ (in-phase) and zero when $\theta = \pi/2$ (out-of-phase). This was illustrated in Figures 8-1 and 8-2 (pages 197-198).

To approximate a median value for phasing, θ is set to $\pi/4$. Here, approximately one-half of the time the MTF will be higher and one-half of the time the MTF will be lower. The median sampling MTF is

$$MTF_{MEDIAN}(u_i) \approx \cos\left(\frac{\pi}{2}\frac{u_i}{u_{iS}}\right). \quad (9\text{-}19)$$

An MTF averaged over all phases is represented by

$$MTF_{AVERAGE}(u_i) \approx sinc\left(\frac{u_i}{u_{iS}}\right). \quad (9\text{-}20)$$

Figure 9-5 illustrates the difference between the two equations. Because these are approximations, they may be considered roughly equal over the range of interest (zero to the Nyquist frequency). This MTF is a function of the sampling frequency, $1/d_{CC}$. The distance between detectors changes with angle[12] and this affects the sample scene MTF. For example, at 45° the spacing is $d_{CC} = (d_{CCH}^2 + d_{CCV}^2)^{1/2}$. For mathematical convenience, MTF_{PHASE} is applied separately to vertical and horizontal MTFs.

238 SAMPLING, ALIASING, and DATA FIDELITY

When the sample-scene MTF is used, only an average system performance is predicted. Performance can vary dramatically depending on the target phase with respect to the sampling lattice. For laboratory measurements, it is common practice to "peak-up" the target (in-phase relationship) and then $MTF_{PHASE} = 1$. This is done to obtain repeatable results.

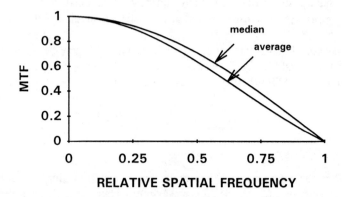

Figure 9-5. Average and median scene-sample phase MTFs normalized to u_i/u_{iS}. This MTF is defined only up to the Nyquist frequency.

9.5. ELECTRONIC ANALOG FILTERS

When filter response is provided as a function of electrical frequency, it is denoted by $H(f_e)$ or $H(f_v)$. When the electrical frequency is transposed to object space or image space, the filter impulse response is labeled as an MTF.

Throughout the text, low-pass filters were used as either anti-alias or reconstruction filters. They also serve another function. Wideband noise is injected by the detector and amplifiers. The noise at the camera's output depends on the electronic bandwidth. By reducing bandwidth (i.e., using a low-pass filter), the output noise is reduced. If $S_{NOISE}(f_e)$ is the noise power spectrum with units of volts/$\sqrt{Hz}$, the total rms noise is

$$V_{NOISE} = \sqrt{\int_0^\infty S_{NOISE}(f_e) H^2(f_e) \, df_e} \, . \tag{9-21}$$

SYSTEM ANALYSIS 239

The signal-to-noise ratio is proportional to

$$SNR(u_o) = k\frac{MTF_{SYSTEM}(u_o)\,O(u_o)}{V_{NOISE}}.\qquad(9\text{-}22)$$

The signal is affected by the optics, detector, and electronic response whereas noise is affected only by the electronics. Appropriately designed electronic filters will not degrade the signal but will minimize the total noise[13] and thereby maximize the SNR.

9.5.1. LOW-PASS FILTER

The ideal low-pass filter is not physically realizable so various compromises are made as part of the design. Figure 9-6 illustrates a tolerance scheme often employed in filter design. A transition region allows a smooth progression from the pass band to the stop band. In the pass band and stop band, some ripple may exist. As the transition bandwidth becomes narrower or as the ripple decreases, filter complexity increases. δ_1 affects in-band amplitudes. For anti-alias filters, δ_2 specifies the amount of aliased signal. For reconstruction filters, it specifies the amount of replicated spectra that will appear in the reconstructed analog signal. Interpolation algorithms can also be used as reconstruction filters. These filters were discussed in Section 6.5., *Interpolation Algorithms* (page 144).

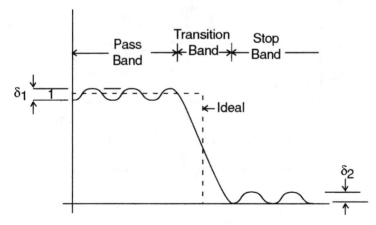

Figure 9-6. A practical filter is designed with allowable tolerances. As $\delta_1 \to 0$, $\delta_2 \to 0$, and the transition region width approaches zero, it becomes an ideal filter.

240 SAMPLING, ALIASING, and DATA FIDELITY

Ideally, $H(f_e)$ and $H(f_v)$ are filters whose MTF is unity up to the array Nyquist frequency and then drops to zero. This filter maximizes the SNR by attenuating out-of-band amplifier noise and passing the signal without attenuation. The ideal filter is unrealizable but can be approximated by N^{th}-order Butterworth or Chebyshev filters. The Butterworth filters are:

$$H_{LOWPASS}(f_e) = \frac{1}{\sqrt{1 + \left(\frac{f_e}{f_{e3dB}}\right)^{2N}}} \quad (9\text{-}23)$$

or

$$H_{LOWPASS}(f_v) = \frac{1}{\sqrt{1 + \left(\frac{f_v}{f_{v3dB}}\right)^{2N}}}, \quad (9\text{-}24)$$

where f_{e3dB} and f_{v3dB} are the frequencies at which the power is one-half or the amplitude is 0.707. As $N \to \infty$, $H_{LOWPASS}(f_e)$ and $H_{LOWPASS}(f_v)$ approach the ideal filter response with the cutoff frequency of f_{e3dB} and f_{v3dB}, respectively (Figure 9-7).

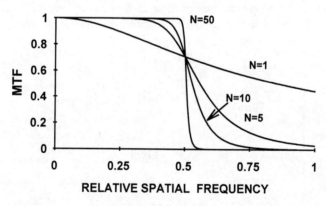

Figure 9-7. Butterworth filters for N = 1, 5, 10, and 50. $f_{e3dB} = 0.5$.

The Nth-order Chebyshev filter MTF is

$$H_{CHEBYCHEV}(f_V) = \frac{1}{\sqrt{1 + (A^2 - 1) C_n^2(f_V)}}. \quad (9\text{-}25)$$

The recursive formulas are
$C_o = 1$
$C_1 = f_V/f_{vC}$
$C_2 = 2(f_V/f_{vC})^2 - 1$
$\overline{} \quad \overline{}$
$C_n = 2(f_V/f_{vC})C_{n-1} - C_{n-2}$,

where f_{vC} is the filter cutoff. As a reconstruction filter, f_{vC} should be equal to the Nyquist frequency. The ripple in dB is Ripple$_{dB}$=20 log(A) or

$$A = 10^{\frac{Ripple}{20}}. \quad (9\text{-}26)$$

Figure 9-8 illustrates a 5th-order Chebyshev filter with 1 dB of ripple. Filters with ripples of 1 dB do not provide adequate amplitude response over the regions of interest.

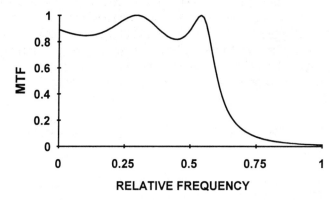

Figure 9-8. Fifth-order Chebyshev filters with the detector MTF. One dB ripple significantly affects the system amplitude response. $f_{vC} = 0.5$.

For analog video output, A low-pass filter is used as a reconstruction filter. This filter affects only the horizontal signal. While the video bandwidth has been standardized (NTSC, PAL, and SECAM is 4.2, 5.5, and 6 MHz, respectively), the filter design has not. Thus, different cameras may have different

242 SAMPLING, ALIASING, and DATA FIDELITY

reconstruction filters and therefore different horizontal MTFs even though they are built to the same video standard. No reconstruction filter is used in the vertical direction. Here, the display and HVS provide the reconstruction. As a result of different detector sizes (d_V and d_H) and (horizontal) analog filters, the MTFs in the vertical and horizontal directions can be quite different.

9.5.2. BOOST

MTF degradation caused by the various subsystems can be partially compensated with electronic boost filters. Boost filters also amplify noise so that the signal-to-noise ratio may degrade. For systems that are contrast limited, boost may improve image quality. However, for noisy images, the advantages of boost are less obvious. As a result, these filters are used only in high contrast situations (typical of consumer applications) and are not used in scientific applications where low signal-to-noise situations are often encountered. Historically, each subsystem was called an "aperture." Therefore, boost provided "aperture correction." This should not be confused with the optics MTF where the optical diameter is the entrance aperture. "Aperture correction" has been used[14] in video for years.

The boost amplifier can either be an analog or digital circuit whose peaking compensates for any specified MTF roll-off. The MTF of a boost filter, by definition, exceeds one over a limited spatial frequency range. When used with all the other subsystem MTFs, the resultant MTF_{SYSTEM} is typically less than one for all spatial frequencies. Excessive boost can cause ringing at sharp edges. The boost algorithm should enhance signal amplitudes with frequencies below the Nyquist because reconstruction usually limits the response to the Nyquist frequency.

A variety of boost circuits are available.[15] One example is the tuned circuit response is

$$H_{BOOST}(f_v) = \frac{1}{\sqrt{\left(1 - \left(\frac{f_v}{f_{BOOST}}\right)^2\right)^2 + \left(\frac{f_v}{Qf_{BOOST}}\right)^2}}, \qquad (9\text{-}27)$$

where Q is the quality factor and equal to boost amplitude when $f_v = f_{BOOST}$.

SYSTEM ANALYSIS 243

Although illustrated in the video domain, boost can be placed in any part of the circuit where a serial stream of data exists. By appropriate selection of coefficients, digital filters can also provide boost. MTF_{BOOST} can also approximate the inverse of MTF_{SYSTEM} (without boost). The combination of MTF_{BOOST} and MTF_{SYSTEM} can provide unity MTF over spatial frequencies of interest. This suggests that the reproduced image will precisely match the target in every spatial detail (Figure 9-9).

Boost also increases aliased signal. In Figure 9-10, the aliased signal is modified by a low-pass filter (such as a reconstruction filter). The amount of boost selected is a tradeoff between boost amplitude and the amount of aliasing tolerated. Aliased signal is scene dependent and therefore the amount of boost desired also becomes scene dependent. This increase in aliased signal limits the amount of image sharpening possible.

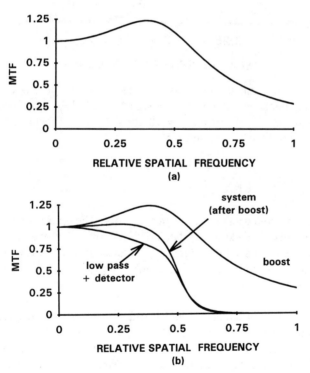

Figure 9-9. (a) Boost filter when $f_{BOOST} = 0.5$ and $Q = 1.1$ and (b) effect on the system response. The system (before boost) includes $MTF_{DETECTOR}$ and a low-pass Butterworth filter with $N = 10$. This boost provides nearly a flat response up to f_N. $f_S = 1$.

244 SAMPLING, ALIASING, and DATA FIDELITY

The unsharp mask[16] is used to emphasize high frequencies and, therefore, is a boost filter. The scene is first low-pass filtered. This image is subtracted from the scene to leave the high frequency components. The difference image is boosted by a gain G and then added to the original image:

$$MTF_{UNSHARP}(u,v) = [1 + G(1 - MTF_{LOWPASS}(u,v))] \ . \qquad (9\text{-}28)$$

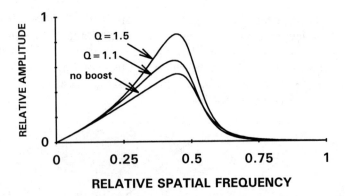

Figure 9-10. The amount of aliased signal depends on the boost frequency and boost amplitude. Aliased signal with $MTF_{DETECTOR}$ and a low-pass Butterworth filter with N = 10 for a 100% fill-factor array. $f_{BOOST} = 0.5$ with Q = 0, 1.1, and 1.5.

9.6. DIGITAL FILTERS

Digital filters process data that reside in a memory. The units assigned to the filter match the units assigned to the data arrays. With one-to-one mapping of pixels to datels, the filter sampling frequency is the same as the array sampling frequency. With this mapping, each filter coefficient processes one pixel value. Digital filters can be two-dimensional.

There are two general classes of digital filters[17]: infinite impulse response (IIR) and finite impulse response (FIR). Both have advantages and disadvantages. The FIR has a linear phase shift whereas the IIR does not. IIR filters tend to have excellent amplitude response whereas FIR filters tend to have more ripple. FIR filters are typically symmetrical in that the weightings are symmetrical about the center sample. They are also the easiest to implement in hardware or software.

SYSTEM ANALYSIS 245

Figure 9-11 illustrates two FIR filters. The digital filter design software provides the coefficients, A_i. The central data point is replaced by the digital filter coefficients as they operate on the neighboring data points. The filter is then moved one data point and the process is repeated until the entire data set has been operated on. Edge effects exist with any digital filter. The filter illustrated in Figure 9-11a requires seven inputs before a valid output can be achieved. At the very beginning of the data set, there are insufficient data points to have a valid output at data point 1, 2, or 3. The user must be aware of edge effects at both the beginning and the end of his data record. In effect, this states that edges cannot be filtered. The following MTF equations are only valid where there are no edge effects. For FIR filters where the multiplicative factors (weightings) are symmetrical about the center, the filter is mathematically represented by a cosine series (sometimes called a cosine filter).

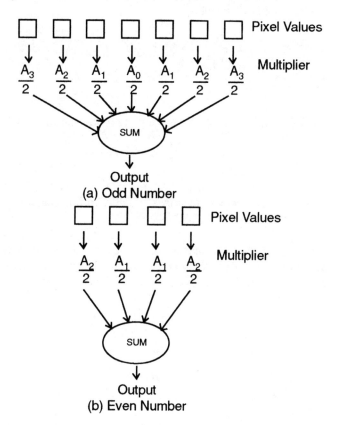

Figure 9-11. Symmetrical digital filters. (a) 7-tap (odd number) filter and (b) 4-tap (even number) filter.

246 SAMPLING, ALIASING, and DATA FIDELITY

Digital filters provide any variety of pass bands to modify the frequency features of a digital data array. Two-dimensional filters are considered separable (i.e., the vertical and horizontal operations are independent). Digital filter response is symmetrical about the Nyquist frequency and repeats at multiples of the sampling frequency. With its frequency response transposed to electrical frequency, an FIR filter with an odd number of samples (also called taps) provides

$$H_{DFILTER}(f_e) = \left| \sum_{k=0}^{\frac{N-1}{2}} A_k \cos\left(\frac{2\pi k f_e}{f_{eS}}\right) \right| . \quad (9\text{-}29)$$

For an even number of samples,

$$H_{DFILTER}(f_e) = \left| A_o + \sum_{k=1}^{\frac{N}{2}} A_k \cos\left(\frac{2\pi (k-1) f_e}{f_{eS}}\right) \right| . \quad (9\text{-}30)$$

The sum of the coefficients should equal unity so that the $H_{DFILTER}(f_e = 0)$ is one:

$$\sum A_k = 1 . \quad (9\text{-}31)$$

Although the above equations provide the filter response in closed form, the response of a real filter is limited by the ability to implement the coefficients. The smallest increment is the LSB.

The averaging filter was introduced in Section 6.2., *Decimation* (page 136) as a simple low-pass filter. With an averaging filter, all the multipliers shown in Figure 9-11 are equal. When N_{AVE} datels are averaged, the equivalent MTF is

$$H_{AVE}(f_e) = \left| \frac{\sin\left(N_{AVE} \pi \frac{f_e}{f_{eS}}\right)}{N_{AVE} \sin\left(\pi \frac{f_e}{f_{eS}}\right)} \right| . \quad (9\text{-}32)$$

The first zero of this function occurs at f_{eS}/N_{AVE}. Figure 9-12 illustrates the averaging effects for several different values of N_{AVE}.

By appropriate selection of digital filter coefficients, peaking can be created. Figure 9-13 illustrates a 7-tap filter that provides peaking.

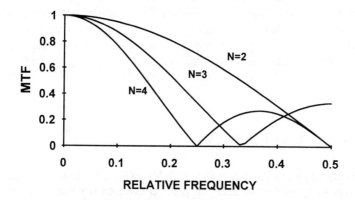

Figure 9-12. MTF of averaging filters normalized to f_e/f_{eS} The filter response is symmetrical about f_N ($f_N = 0.5$).

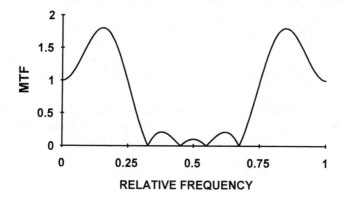

Figure 9-13. Symmetric (cosine) 7-tap digital peaking filter normalized to f_e/f_{eS}. $A_0 = 0.7609$, $A_1 = 0.9115$, $A_2 = -0.2100$, and $A_3 = -0.4624$. The filter response is symmetrical about the Nyquist frequency and repeats at multiples of the sampling frequency.

9.7. DISPLAY

For most electronic imaging systems, the display medium acts as a reconstruction filter. The display was introduced in Section 7.3., *Reconstruction by the Display Medium* (page 170). Displays do not completely attenuate the replicated frequency amplitudes created by the sampling process. The HVS further reduces the sampling remnants so that the perceived image appears continuous with minimal sampling artifacts. Therefore, displays cannot be analyzed in isolation but must be considered as part of the display/observer system.

9.7.1. CRT-BASED DISPLAY

The display MTF is a composite that includes both the internal amplifier and the CRT responses. Implicit in the MTF is the conversion from input voltage to output display brightness. Although not explicitly stated, the equation implies radial symmetry and the MTF is the same in both the vertical and horizontal directions. That is, the MTF is considered separable and $MTF_{CRT}(u_d, v_d) = MTF_{CRT}(u_d)MTF_{CRT}(v_d)$ and $MTF_{CRT}(v_d) = MTF_{CRT}(u_d)$. While this may not be precisely true, the assumption is adequate for most calculations.

It is reasonable to assume that the resultant spot is Gaussian:

$$MTF_{CRT}(u_d) \approx e^{-2\pi^2 \sigma_{SPOT}^2 u_d^2} = e^{-2\pi^2 \left(\frac{S}{2.35} u_d\right)^2}, \qquad (9\text{-}33)$$

where σ_{SPOT} is the standard deviation of the Gaussian beam profile, and S is the spot diameter at full-width, half-maximum. These parameters must have units of millimeters because the frequency has units of cycles/mm. S is

$$S = \sqrt{8\ln(2)}\ \sigma_{SPOT} = 2.35\ \sigma_{SPOT}. \qquad (9\text{-}34)$$

If the spot size is not specified, then σ_{SPOT} can be estimated from the TV limiting resolution. With the TV limiting resolution test, an observer views a wedge pattern and selects that spatial frequency at which the converging bar pattern can no longer be seen. The display industry has standardized the bar spacing to $2.35\sigma_{SPOT}$.

If N_{TV}/PH is the TV limiting resolution and N_{TV} lines are displayed on a monitor, then $H_{MONITOR} \approx 2.35\sigma_{SPOT}N_{TV}$. When transposed to image space,

$$MTF_{CRT}(u_i) \approx e^{-2\pi^2\left(\frac{VFOVfl}{2.35N_{TV}}u_i\right)^2} . \qquad (9\text{-}35)$$

For staring arrays with N_V displayed vertical lines,

$$MTF_{CRT}(u_i) = e^{-2\pi^2\left(\frac{(N_V-1)d_{CCV}+d_V}{2.35N_{TV}}u_i\right)^2} . \qquad (9\text{-}36)$$

Many consumer CCD cameras match the number of vertical detectors to the number of active video lines. With consumer televisions, $N_{TV} \approx 0.7N_V$. Thus, the video standard uniquely defines the CRT MTF. For 100% fill-factor arrays,

$$MTF_{CRT}(u_i) \approx e^{-2\pi^2\left(\frac{d_{CCV}}{1.645}u_i\right)^2} . \qquad (9\text{-}37)$$

Figure 9-14 illustrates the MTFs for a 100% fill-factor array ($d_{CCV} = d_V$). At Nyquist frequency ($u_N = 1/2d_{CCV}$), $MTF_{CRT} = 0.161$. That is, the "standard" design effectively reduces sampling effects by attenuating the amplitudes above f_N. However, the in-band amplitudes are unavoidably attenuated also.

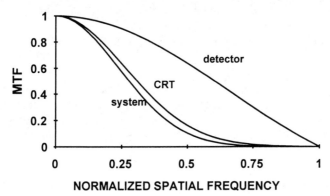

Figure 9-14. MTF of a "standard" raster-scanned CRT and 100% fill-factor array. The number of raster lines is equal to the number of detector elements in the vertical direction. The CRT response significantly reduces the in-band signal amplitudes. The optical MTF is considered unity over these frequencies: $MTF_{SYSTEM} \approx MTF_{CRT} \, MTF_{DETECTOR}$.

9.7.2. FLAT PANEL DISPLAY

Flat panel displays consist of discrete elements. If the emitting area is rectangular, it provides a sinc response:

$$MTF_{FLAT}(u_d, v_d) = sinc(d_{H-FP} u_d) \, sinc(d_{V-FP} v_d) \, , \qquad (9\text{-}38)$$

where $d_{H-FP} \times d_{V-FP}$ is the size of the emitting area. The sinc function does not adequately attenuate sampling artifacts (see Figure 7-26, page 181). Therefore, flat panel displays rely heavily on the observer's response to create a perceived continuous image. The individual elements are purposely made small so that the eye blends them. The eye also significantly reduces in-band signal amplitudes.

9.8. THE OBSERVER

The human visual system MTF (sometimes labeled as HVS-MTF) was previously introduced as a reconstruction filter (see Section 7.2., *The Observer*, page 167). The sine wave response is used as an approximation to the HVS. The sine wave response depends on diffraction by the pupil, aberrations of the lens, finite size of the photoreceptors, ocular tremor, and neural interconnections within the retina and brain. Diffraction and aberrations vary with ambient luminance, monitor brightness, and chromatic composition of the light. Because the retina is composed of rods and cones of varying densities, the location and size of the object with respect to the fovea will significantly affect the sine wave response.

The sine wave response is only an approximation to the true response. Furthermore, the overall population exhibits large variations in response. Any MTF approximation used for the eye therefore is only a crude approximation and probably represents the largest uncertainty in the overall MTF analysis approach.

The contrast sensitivity is the inverse of the threshold modulation curve illustrated in Figure 7-11 (page 167). When normalized to one, the contrast sensitivity or sine wave response is called the HVS-MTF. By convention, the MTF is defined to be one at zero spatial frequency and to be a monotonically decreasing function. The way to include the inhibitory response as an MTF component is not clear.

Various researchers have taken the inverse of the threshold modulation, normalized it, and labeled[18-21] it as the observer MTF. Although the eye operates in a logarithmic fashion, the MTF is plotted on a linear scale. As a reconstruction filter, only the high spatial frequency response is important. An approximation (Figure 9-15) that ignores inhibition is[22]:

$$MTF_{OBSERVER}(u_{eye}) = e^{-\Gamma \frac{u_{eye}}{17.45}}, \quad (9\text{-}39)$$

where Γ is a light-level dependent eye response factor that was presented in tabular form[23] by Ratches et al. Figure 9-16 illustrates the third-order polynomial fit given by

$$\Gamma = 1.444 - 0.344\log(B) + 0.0395\log^2(B) + 0.00197\log^3(B), \quad (9\text{-}40)$$

where B is the monitor brightness in foot-lamberts.

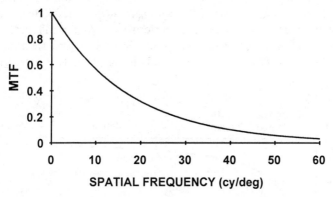

Figure 9-15. Representative $MTF_{HVS}(u_{eye})$ as a function of u_{eye}. $\Gamma = 1$.

Figure 9-16. Gamma.

9.9. SYSTEM MTF

When viewing a point source, diffraction-limited optics produces a diffraction pattern that consists of a bull's-eye pattern with concentric rings. The center is the Airy disk and its diameter is

$$d_{AIRY} = 2.44 \frac{\lambda fl}{D_o} = 2.44 \lambda F . \qquad (9\text{-}41)$$

Assume that the detector array has square detectors with 100% fill-factor ($d_H = d_V = d_{CCV} = d_{CCH} = d$). When d_{AIRY} is larger than the detector size, the system is said to be optics-limited. Here, changes in the optics MTF (e.g., D_o or fl) significantly affect MTF_{SYSTEM}. If $d_{AIRY} < d$, the system is detector-limited and changes in the detector size affect MTF_{SYSTEM} at u_N. Figure 9-17 illustrates $MTF_{OPTICS}MTF_{DETECTOR}$ at Nyquist frequency as a function of $d/F\lambda$. It is the ratio of the optical cutoff to the detector cutoff: $u_{iC}/u_{iD} = d/F\lambda$. The MTF at Nyquist frequency is often used as a measure of performance. As the MTF increases, image quality increases. Unavoidably, as the MTF increases, aliasing also increases. This tradeoff is further discussed in Chapter 11, *Image Quality Metrics*.

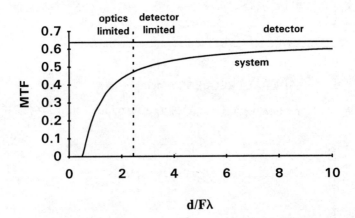

Figure 9-17. $MTF_{OPTICS}MTF_{DETECTOR}$ at Nyquist frequency as a function of $d/F\lambda$. The fill-factor is 100% ($d = d_{CC}$). The vertical line indicates where the Airy disk diameter is equal to the detector size. The best MTF that can be achieved occurs when MTF_{OPTICS} is negligible compared to $MTF_{DETECTOR}$. Equivalently, $MTF_{OPTICS}MTF_{DETECTOR} \approx MTF_{DETECTOR} = 0.637$.

For visible systems, $\lambda \approx 0.5\,\mu m$ and the transition from the optics-limited to detector-limited case occurs when $F = d/1.22$. Because d is typically 10 μm, then F must be greater than 8.2 to enter the optics-limited condition (Figure 9-18). This may occur with telephoto lenses. For infrared systems operating in the mid-wave region (3-5 μm), $\lambda_{AVE} \approx 4.0\,\mu m$ and the transition occurs at $F = d/9.76$. For mid-wave infrared detectors, $d \approx 40$ μm and the system becomes optics-limited when $F > 4.1$. But, for long-wave infrared ($\lambda_{AVE} \approx 10.0\,\mu m$), the transition occurs at $F = d/24.4$. Then the system becomes optics-limited when $F > 1.6$. Thus, visible systems tend to be detector-limited and long-wave infrared systems tend to be optics-limited.

When $d/F\lambda = 0.5$, there are two detector samples per highest optical spatial frequency. The MTF at u_N is zero because u_N is equal to the optical cutoff. This represents 4.88 samples across the Airy disk. With an ideal reconstruction filter, the Airy disk can be accurately reconstructed and the sampling theorem is satisfied. Some degradation exists due to the integration capability provided by the detector (its sinc function). Image enhancement or boost can offset this degradation so that the Airy disk can be completely recovered. However, referring to Figure 9-18, unless the f-number is extremely large, this situation cannot exist with typical detector sizes currently available. That is, aliasing will occur with nearly all electronic imaging systems.

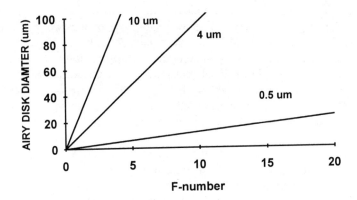

Figure 9-18. Airy disk diameter as a function of f-number for three different wavelengths. When the $d < d_{AIRY}$, the system is optics-limited. When $d > d_{AIRY}$, the system is detector-limited.

254 SAMPLING, ALIASING, and DATA FIDELITY

Usually the detector size is fixed. The parameter, $d/F\lambda$, is

$$\frac{d}{F\lambda} = \left(\frac{d}{fl}\right)\frac{D_o}{\lambda} = DAS\frac{D_o}{\lambda}. \tag{9-42}$$

The output voltage (for square detectors) is given by the simple camera formula:

$$v_{CAMERA} = \frac{1}{4}\left(\frac{dD_o}{fl}\right)^2 \int_{\lambda_1}^{\lambda_2} R(\lambda) M_{illum}(\lambda) d\lambda, \tag{9-43}$$

where $R(\lambda)$ is the spectral responsivity and $M_{illum}(\lambda)$ is the illumination. Decreasing the focal length, increases the MTF at u_N. It also increases the camera output (improves sensitivity). But decreasing the focal length increases the scene spatial frequencies relative to the detector response. Thus, a tradeoff exists between the spatial frequencies that can be reproduced and sensitivity.

While parametric analyses are convenient, absolute values are lost. The MTF at Nyquist frequency is important. However, it is even more important to know the relationship between the scene spatial content and the actual value of the Nyquist frequency. In object space, as the focal length increases, the absolute value of the Nyquist frequency also increases.

It is instructive to see the effects of $d/F\lambda$ on computer generated imagery. Details of the simulation are provided in the next section. Figure 9-19a illustrates the original scene and Figure 9-19b provides the two-dimensional transform. When $d/F\lambda = 0.5$, $MTF(u_N) = 0$ and the optics have low-pass filtered the scene. Here, no aliasing can occur. Figure 9-20a provides the MTFs. Figure 9-20b provides the two-dimensional transform which includes a Butterworth reconstruction filter where $N = 10$. Figure 9-20c illustrates the resultant imagery. While the low-pass filtering action of the optics removed all aliasing, it also significantly attenuated the in-band frequency amplitudes.

Figures 9-21 and 9-22 illustrate the cases where $d/F\lambda = 2.44$ and 10, respectively. Note that the MTFs illustrated in Figures 9-21a and 9-22a do not include the reconstruction filter. The transforms and imagery include a 10th-order Butterworth reconstruction filter. MTFs with the reconstruction filter are illustrated in Section 9.10.6., *Reconstruction Filter*. Careful examination of Figure 9-22c will reveal that the edges are sharper compared to Figure 9-21c but aliasing has increased. However, the effect is not dramatic when using a high order Butterworth reconstruction filter. At Nyquist frequency, $MTF_{OPTICS}MTF_{DETECTOR}$ increased from 0.472 to 0.592. In terms of imagery, whether $d/F\lambda = 10$ is better than $d/F\lambda = 2.44$ depends on the scene spectral content and the reconstruction filter used. The amount of aliased signal is scene dependent and the amount of aliasing accepted depends on personal preference.

Figure 9-19. (Upper) Image and (lower) two-dimensional transform. As the spatial frequency increases, the magnitude of $F(u,v)$ drops. By using a logarithmic gray scale, $F(u,v)$ becomes visible for all frequencies in this figure. Created by the *System Image Analyzer* software.[24]

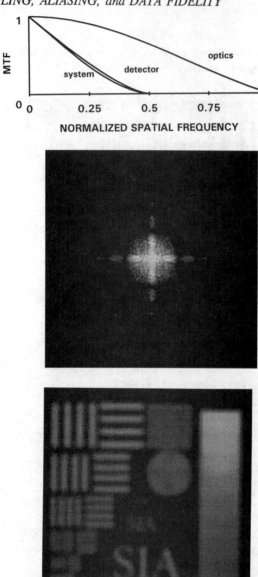

Figure 9-20. (Upper) MTF_{OPTICS} and $MTF_{DETECTOR}$ when $d/F\lambda = 0.5$. (Middle) Two-dimensional transform including a Butterworth filter where $N = 10$. Gray scale is logarithmic. (Lower) Reconstructed image. The optics has significantly blurred the image and no aliasing exists. Created by the *System Image Analyzer* software.[24]

SYSTEM ANALYSIS 257

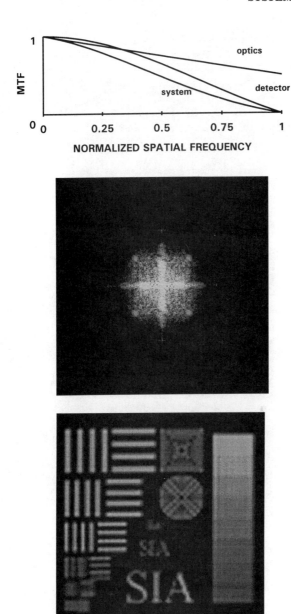

Figure 9-21. (Upper) MTF_{OPTICS} and $MTF_{DETECTOR}$ when $d/F\lambda = 2.44$. (Middle) Two-dimensional transform including a Butterworth filter where $N = 10$. Gray scale is logarithmic. (Lower) Reconstructed image.

Figure 9-22. (Upper) MTF$_{OPTICS}$ and MTF$_{DETECTOR}$ when $d/F\lambda = 10$. (Middle) Two-dimensional transform including a Butterworth filter where $N = 10$. Gray scale is logarithmic. (Lower) Reconstructed image.

9.10. SIMULATION OF SUBSYSTEM MTFs

Using the display and HVS as reconstruction filters masks most subsystem MTFs. The starting point to study MTFs is acquiring a high resolution image. This image can be the output of another sensor or may be computer generated. It becomes the scenels for the simulation. With several scenels per pixel, the scene frequency content is shifted down relative to the display and HVS frequency responses. Generally, if there are 16 scenels per pixel (4 × 4:1), then the display and HVS MTFs do not significantly degrade the imagery. In addition, with 16 or more scenels per pixel, aliasing is easily demonstrated.

With a computer simulation, it is important to select the appropriate parameters so that the desired results are achieved. A DFT of the simulated image provides its frequency components. This transformed image is multiplied by $MTF_{OPTICS}MTF_{DETECTOR}$, has replicated frequency spectra added, is multiplied by the remaining MTFs, and then inverse transformed to obtain the simulated image. Usually the number of scenels, datels, and disels is the same.

Because the DFT processing time increases as the scene size increases, subarray analysis offers moderate transform time while providing the desired results. The analyzed field-of-view is a fraction of the actual field-of-view.

9.10.1. COMPUTER GENERATED IMAGERY

Computer algorithms assign values to individual scenels. That is, the bar widths will be an integer number of scenels: 1, 2, ⋯. This limits the number of frequencies in the scene. The aliased spectrum cannot exist because no computer generated target feature will be less than one scenel wide. Selecting sixteen scenels per pixel minimizes phasing effects. Figures 8-22 though 8-24 (pages 216-217) illustrated that as the target frequency decreases, phasing effects also decrease. Sixteen scenels/pixel also minimizes the fact that certain frequencies cannot exist.

9.10.2. HIGH RESOLUTION IMAGERY

If the image comes from another camera, it already contains aliased components created by that camera's detector array. It is necessary to divide $I_{HI-RES}(u,v)$ by the camera's MTF to obtain an estimate of the original imagery. For many systems, the optical and detector MTFs dominate.

260 SAMPLING, ALIASING, and DATA FIDELITY

The data represent a high-resolution image whose frequency components are

$$I_{HI\text{-}RES}(u,v) = MTF_{DETECTOR\text{-}HI\text{-}RES}(u,v)\, MTF_{OPTICS\text{-}HI\text{-}RES}(u,v)\, I(u,v) \ . \qquad (9\text{-}44)$$

The frequency components of the simulated image are

$$I_{SIMULATED}(u,v) = \frac{I_{HI\text{-}RES}(u,v)}{MTF_{DETECTOR\text{-}HI\text{-}RES}(u,v)\, MTF_{OPTICS\text{-}HI\text{-}RES}(u,v)} \ . \qquad (9\text{-}45)$$

If scanning a printed image, then the printer and scanner MTFs should be included. While the division accentuates the aliased signal, the aliased signal will be reduced by the optical and detector MTFs of the simulation. The datels representing $i_{HI\text{-}RES}(x,y)$ become the scenels for $i_{SIMULATED}(x,y)$. For subarray analysis, $I_{SIMULATED}(u,v)$ is used.

To see an image that would be created by the camera being studied, $I_{SIMULATED}(u,v)$ is first multiplied by MTF_{OPTICS} and then by $MTF_{DETECTOR}$. Next, frequency spectra are replicated to simulate aliasing. As appropriate, additional MTFs may be employed. This spectrum is downsampled by selecting every $(N_{DETECTOR}/N_{SCENEL})$-th datel and discarding the remaining. To avoid additional aliasing, a low-pass filter must be used (see Section 6.2., *Decimation*, page 136). The resultant simulates an image that would have been obtained from the camera. The display and HVS provide the reconstruction filters.

9.10.3. ALIASING

Without frequency replications, aliasing will not be present in the simulated image. Replications occur only when specific conditions are met. Let a scene consists of $N_{SCENEL\text{-}H} \times N_{SCENEL\text{-}V}$ scenels and let there be $N_H \times N_V$ detectors. When represented in object space, the discrete Fourier transform limits the simulated scene frequencies to

$$u_{SCENEL\text{-}N} = \frac{N_{SCENEL\text{-}H}}{2\,HFOV} \quad and \quad v_{SCENEL\text{-}N} = \frac{N_{SCENEL\text{-}V}}{2\,VFOV} \quad \frac{cycles}{mrad} \ . \qquad (9\text{-}46)$$

SYSTEM ANALYSIS 261

For computer-generated imagery, the highest frequencies present are $u_{SCENEL-N}$ and $v_{SCENEL-N}$. That is, the imagery is band-limited to these values. In object space, the detector array Nyquist frequency is

$$u_{oN} = \frac{N_H}{2HFOV} \quad and \quad v_{oN} = \frac{N_V}{2VFOV} \quad \frac{cycles}{mrad}. \qquad (9\text{-}47)$$

To simulate aliasing, $u_{oN} < u_{SCENEL-N}$ and $v_{oN} < v_{SCENEL-N}$. Equivalently, $N_H < N_{SCENEL-H}$ and $N_V < N_{SCENEL-V}$. When $N_H = N_{SCENEL-H}/4$ and $N_V = N_{SCENEL-V}/4$, two replications are possible in the two-dimensional frequency map created by the DFT (Figure 9-23).

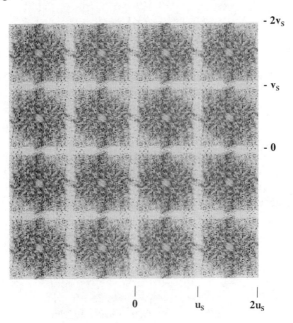

Figure 9-23. Replicated spectrum of Figure 9-19b with the detector sinc function included. with $N_H = N_V = N_{SCENEL}/4$, two replicated spectra are created. The number of scenels is 256 × 256 and the number of pixels is 64 × 64. Aliasing appears as overlapping spectra. Gray scale is logarithmic.

9.10.4. NUMBER OF DETECTORS

Each pixel is mapped to a datel. With one-to-one mapping, one disel is allocated to a detector. For example, if the number of scenels is 256 × 256 and the number of detectors is 64 × 64, there will be 256 × 256 disels that contain 64 × 64 equally spaced pixels.

Rather than start with a fixed number of detectors and then calculating the number of scenels required, it is easier to do the reverse. Because most DFT algorithms operate on powers-of-two, a modest size is selected such as 256 × 256. This number of horizontal scenels represent

$$N_{SCENEL-H} = (N_H - 1)N_{PITCH} + N_{DET} , \qquad (9\text{-}48)$$

where N_{PITCH} is the number of scenels spanning the detector pitch and N_{DET} is the number of scenels spanning the active area of a detector. Let the linear fill-factor be

$$L_{RATIO} = \frac{N_{DET}}{N_{PITCH}} . \qquad (9\text{-}49)$$

For square detectors, $(L_{RATIO})^2$ is the fill-factor. Substituting provides

$$N_H = L_{RATIO}\left(\frac{N_{SCENEL}}{N_{DET}}\right) - L_{RATIO} + 1 . \qquad (9\text{-}50)$$

For example, consider an image that is 256 × 256 (N_{SCENEL} = 256). Let L_{RATIO} = 10/12 = 0.8333. With N_{DET} = 4, two frequency replications are possible. Then N_H = 53.5 detectors. For this example, the subarray should contain 53.5 pixels. While a fractional number of detectors are not physically realizable, most computer programs do not place any restrictions on input values.[24]

SYSTEM ANALYSIS 263

9.10.5. FREQUENCY SCALING

Subarray analysis reduces the horizontal and vertical fields-of-view. This changes the scaling factors for the electronic filter, video filter, monitor, and observer MTFs. To maintain the same relative MTFs:

1. ANALOG ELECTRONICS: Equation 9-5 (page 231) scales electronic frequency into the image space domain with a factor of $HFOV/t_{H-LINE}$. Changing the HFOV requires changing t_{H-LINE} in the same proportion as the HFOV.

2. VIDEO ELECTRONICS: Equation 9-8 (page 232) scales video frequency into the image space domain with a factor of $HFOV/t_{VIDEO-LINE}$. If the HFOV is reduced then the active line time must be reduced in the same proportion to maintain the same filter cutoff frequency.

3. CRT-BASED MONITOR: Equation 9-35 (page 249) scales display frequency into the image space domain with a factor of $VFOV/N_{TV}$. If the VFOV is reduced, the monitor resolution, N_{TV}, must be reduced in the same proportion.

4. OBSERVER: Equations 9-10 and 9-14 (pages 232-233) scale observer frequency into the image space domain with a factor of $HFOV/W_{MONITOR}$. If the HFOV is reduced then the monitor width must be reduced in the same proportion to maintain the same viewing distance. Recall that the observer is not normally used when creating imagery.

9.10.6. RECONSTRUCTION FILTERS

The inverse transform of Figure 9-23 (with no reconstruction filter) provides 64 × 64 finite data that represent detector locations. As digital data, these are impulses but become visible due to the reconstruction filter in the laser printer. That is, the computer sent a 256 × 256 data array to the printer and the low-pass filtering action turned the data into visible ink blobs (Figure 9-24). Because the pixels are separated, their blur spots do not overlap and they can be seen in the displayed image. Equivalently, the reconstruction process included the sampling frequencies so that the sampling process is evident in the displayed image.

264 SAMPLING, ALIASING, and DATA FIDELITY

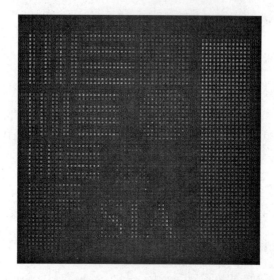

Figure 9-24. Inverse transform of Figure 9-23. No reconstruction filter was used and, therefore, the discrete nature of the data is evident.

Figures 9-25 through 9-27 illustrate how a Nth order low-pass reconstruction filter affects imagery. When scaled into image space, the reconstruction filter is of the form

$$MTF_{RECON}(u_i,v_i) = \frac{1}{\sqrt{1 + \left(\frac{u_i}{u_{iN}}\right)^{2N}}} \frac{1}{\sqrt{1 + \left(\frac{v_i}{v_{iN}}\right)^{2N}}}, \qquad (9\text{-}51)$$

where N is 1, 10, and 100 in Figures 9-25, 9-26, and 9-27, respectively. Figures 9-25a, -26a, and -27a provide the MTFs where $d/F\lambda = 10$. The optics MTF is not shown to highlight the effect of the reconstruction filter. Only the nonaliased component of the detector MTF is shown to illustrate the importance of reconstruction filters. Figures 9-25b, -26b, and -27b provide the two-dimensional transforms after the reconstruction filter. These figures include aliasing caused by the detector. Figures 9-25c, -26c, and -27c provide the inverse transformed image. When N is very large, the filter appears like an ideal filter: unity response up to u_{iN} and zero response above u_{iN}. With 16 disels per pixel, the laser printer low-pass filtering action does not significantly affect the printed imagery. Because the image is large, the HVS MTF does not attenuate the image frequencies at normal viewing distances. As the filter cutoff sharpens, the amount of ringing increases (Gibbs phenomenon).

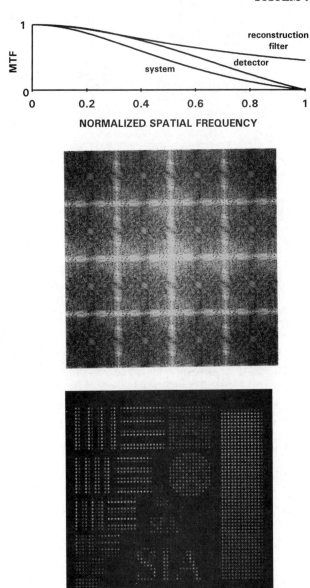

Figure 9-25. Low-pass Butterworth reconstruction filter with $N = 1$. (Upper) One-dimensional MTF as a function of u_i/u_{id}. (Middle) Two-dimensional transform including the reconstruction filter where the gray scale is logarithmic. (Lower) Reconstructed image. The MTF curve does not include aliased components whereas the imagery does.

266 SAMPLING, ALIASING, and DATA FIDELITY

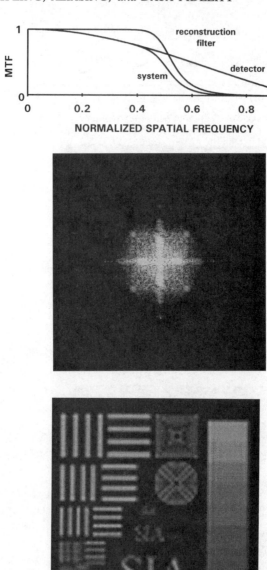

Figure 9-26. Low-pass Butterworth reconstruction filter with N = 10. (Upper) One-dimensional MTF as a function of u_i/u_{id}, (Middle) two-dimensional transform including the reconstruction filter, and (lower) reconstructed image. This is identical to Figure 9-22.

SYSTEM ANALYSIS 267

Figure 9-27. Low-pass Butterworth reconstruction filter with N = 100. (Upper) One-dimensional MTF as a function of u_i/u_{id}. (Middle) Two-dimensional transform including the reconstruction filter where the gray scale is logarithmic. (Lower) Reconstructed image.

268 SAMPLING, ALIASING, and DATA FIDELITY

Figures 9-28 through 9-30 show the effect of various CRT spot sizes. When scaled into image space, the display MTF is modeled by

$$MTF_{CRT}(u_i) \approx e^{-2\pi^2 \left(\frac{VFOV fl}{2.35 N_{TV}} u_i\right)^2}. \qquad (9\text{-}52)$$

When significant frequencies exist above u_{iN}, the sampling lattice is apparent. The sampling lattice disappears when the MTF is zero above u_{iN} (Figure 9-30). However, the in-band frequencies are also attenuated and edges are significantly softened. Figures 9-27c and 9-30c illustrate the tradeoff between filter sharpness and unavoidable attenuation in the pass band.

9.11. DYNAMIC SCENE PROJECTORS

Most test procedures involve human observation of test patterns. Therefore machine vision systems typically cannot be tested by traditional methods. This is particularly true when the imaging system is mounted on a missile. As the missile heads toward the target (range closure), the target size increases. Further, due to flight variations, the target does not remain fixed in the center of the field-of-view. Field testing is expensive and the missile is usually destroyed on impact.

To overcome field test limitations, considerable interest exists in laboratory testing. This has become known as hardware-in-the-loop (HWIL) testing.[25] The computer that creates the image is called a scene generator. This electronic image then is converted into an infrared image by a scene projector (most missile seekers contain an infrared imaging system). These two components are sometimes called a dynamic infrared scene projector (DIRSP). With suitable feedback, range closure and flight variations can be simulated (Figure 9-31). This is a d/c/d system where the missile output is digital. Although the scene generator operates in the digital domain, it acts as a c/d system by simulating the real world. Thus HWIL operation simulates a c/d/c/d system

Key to the design is the number of scenels per pixel. Scenels are expensive. While 16 or more scenels per pixel are desirable, current systems strive for four scenels/pixel. Four is considered[26] *...a tradeoff between cost of providing detailed information and suppression of the more objectionable features of synthetic imagery.* The "objectionable features" are aliasing and edge ambiguity. The methods described in Section 9.10 can be applied to dynamic scene projectors. As with any system, the amount of aliasing tolerated (it always exists) depends on the application.

SYSTEM ANALYSIS 269

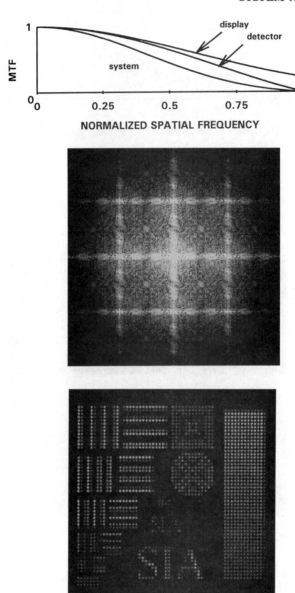

Figure 9-28. CRT display with a small spot size. (Upper) One-dimensional MTF as a function of u_i/u_{id}. (Middle) Two-dimensional transform including the reconstruction filter where the gray scale is logarithmic. (Lower) Reconstructed image. The MTF curve does not include aliased components whereas the imagery does.

270 SAMPLING, ALIASING, and DATA FIDELITY

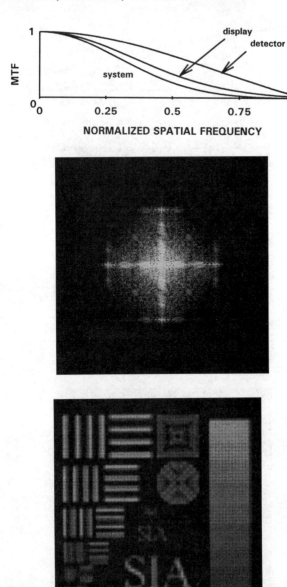

Figure 9-29. CRT display with a moderate spot size. (Upper) One-dimensional MTF as a function of u_i/u_{id}. (Middle) Two-dimensional transform including the reconstruction filter where the gray scale is logarithmic. (Lower) Reconstructed image.

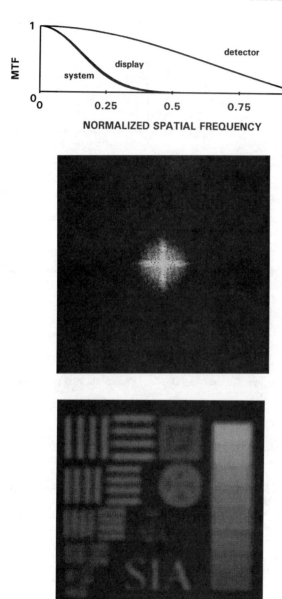

Figure 9-30. CRT display with a large spot size. (Upper) One-dimensional MTF as a function of u_i/u_{id}, (Middle) Two-dimensional transform including the reconstruction filter where the gray scale is logarithmic. (Lower) Reconstructed image.

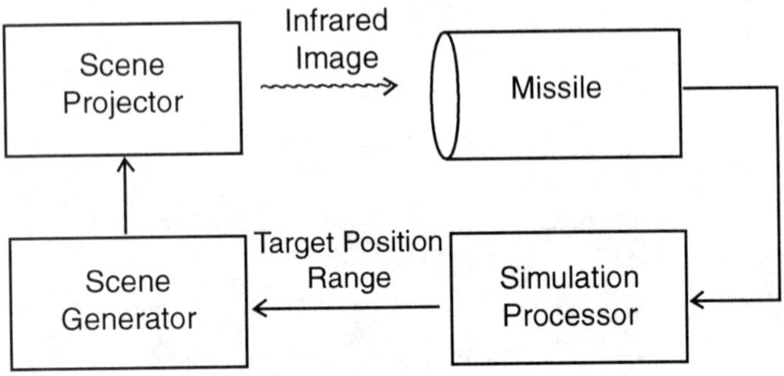

272 SAMPLING, ALIASING, and DATA FIDELITY

Figure 9-31. Typical HWIL setup. The simulation processor is actually a large software program.

9.12. REFERENCES

1. G. C. Holst, *CCD Arrays, Cameras, and Displays*, Chapter 10, JCD Publishing. Winter Park FL (1995).
2. G. C. Holst, *Electro-Optical Imaging System Performance*, Chapters 6 through 11, JCD Publishing. Winter Park FL (1994).
3. G. C. Holst, *Electro-Optical Imaging System Performance*, pp. 106-108, JCD Publishing. Winter Park FL (1994).
4. G. C. Holst, *Electro-Optical Imaging System Performance*, pp. 195-200, JCD Publishing. Winter Park FL (1994).
5. G. C. Holst, *Electro-Optical Imaging System Performance*, pp. 104-105, JCD Publishing. Winter Park FL (1994).
6. S. K. Park and R. A. Schowengerdt, "Image Sampling, Reconstruction and the Effect of Sample-scene Phasing," *Applied Optics*, Vol. 21(17), pp. 3142-3151 (1982).
7. J. C. Feltz and M. A. Karim, "Modulation Transfer Function of Charge-coupled Devices," *Applied Optics*, Vol. 29(5), pp. 717-722 (1990).
8. J. C. Feltz, "Development of the Modulation Transfer Function and Contrast Transfer Function for Discrete Systems, Particularly Charge-coupled Devices," *Optical Engineering*, Vol. 29(8), pp. 893-904 (1990).
9. L. de Luca and G. Cardone, "Modulation Transfer Function Cascade Model for a Sampled IR Imaging System," *Applied Optics*, Vol. 30(13), pp. 1659-1664 (1991).
10. A. Friedenberg, "Comment on the paper 'Development of the Modulation Transfer Function and Contrast Transfer Function for Discrete Systems, Particularly Charge-coupled Devices,'" *Optical Engineering*, Vol. 35(7), pp. 2105-2106 (1996).
11. F. A. Rosell, "Effects of Image Sampling," in *The Fundamentals of Thermal Imaging Systems*, F. Rosell and G. Harvey, eds., page 217, NRL Report 8311, Naval Research Laboratory, Wash D.C. (1979).

12. O. Hadar, A. Dogariu, and G. D. Boreman, "Angular Dependence of Sampling MTF," *Applied Optics*, Vol. 36, pp. 7210-7216 (1997).
13. G. C. Holst, *Electro-Optical Imaging System Performance*, pp. 347-378, JCD Publishing. Winter Park FL (1994).
14. O. H. Shade, Sr., "Image Gradation, Graininess, and Sharpness in Television and Motion Picture Systems," published in four parts in *J. SMPTE*: "Part I: Image Structure and Transfer Characteristics," Vol. 56(2), pp. 137-171 (1951); "Part II: The Grain Structure of Motion Pictures - An Analysis of Deviations and Fluctuations of the Sample Number," Vol. 58(2), pp. 181-222 (1952); "Part III: The Grain Structure of Television Images," Vol. 61(2), pp. 97-164 (1953); "Part IV: Image Analysis in Photographic and Television Systems," Vol. 64(11), pp. 593-617 (1955).
15. G. C. Holst, *Electro-Optical Imaging System Performance*, pp. 144-148, JCD Publishing. Winter Park FL (1994).
16. A. R. Weeks Jr, *Fundamentals of Electronic Image Processing*, pp. 139-144, SPIE Optical Engineering Press, Bellingham WA (1996).
17. A variety of texts exists on digital filter design. See, for example, *Digital Signal Processing*, A. V. Oppenheim and R. W. Schafer, Prentice-Hall, New Jersey (1975) or J. G. Proakis and D. G. Manolakis, *Digital Signal Processing: Principles, Algorithms, and Applications*, 3rd Edition, Prentice-Hall, Upper Saddle NJ (1996).
18. N. Nill, "A Visual Model Weighted Cosine Transform for Image Compression and Quality Measurements," *IEEE Transactions on Communications*, Vol. 33(6), pp. 551-557 (1985).
19. T. J. Schulze, "A Procedure for Calculating the Resolution of Electro-Optical Systems," in *Airborne Reconnaissance XIV*, P A. Henkel, F. R. LaGesse, and W. W. Schurter, eds., SPIE Proceedings Vol. 1342, pp. 317-327, (1990).
20. F. W. Campbell and J. G. Robson, "Application of Fourier Analysis to the Visibility of Gratings," *Journal of Physiology*, Vol. 197, pp. 551-566 (1968).
21. A. N. deJong and S. J. M. Bakker, "Fast and Objective MRTD Measurements," in *Infrared Systems - Design and Testing*, P. R. Hall and J. S. Seeley, eds., SPIE Proceedings Vol. 916, pp. 127-143 (1988).
22. G. H. Kornfeld and W. R. Lawson, "Visual Perception Model," *Journal of the Optical Society of America*, Vol. 61(6), pp. 811-820 (1971).
23. J. Ratches, W. R. Lawson, L. P. Obert, R. J. Bergemann, T. W. Cassidy, and J. M. Swenson, "Night Vision Laboratory Static Performance Model for Thermal Viewing Systems," US Army Electronics Command Report ECOM Report 7043, page 11, Ft. Monmouth, NJ (1975).
24. *System Image Analyzer* is available from JCD Publishing, Winter Park, FL (1995).
25. See, for example, *Technologies for Synthetic Environments: Hardware-in-the Loop Testing*, R. L. Murrer, ed., SPIE Vol. 2741 (1996), or *Technologies for Synthetic Environments: Hardware-in-the Loop Testing II*, R. L. Murrer, ed., SPIE Vol. 3084 (1997).
26. O. M. Williams, M. A. Manzardo, and E. E. Burroughs, "Image Filtering and Sampling in Dynamic Infrared Projection Systems," in *Technologies for Synthetic Environments: Hardware-in-the Loop Testing II*, R. L. Murrer, ed., SPIE Vol. 3084 (1997).

10

SYSTEM RESOLUTION

An overwhelming majority of image quality discussions center on resolution. Resolution has been in use so long that it is thought to be fundamental and that it uniquely determines system performance. It implies something about the smallest target detail that can be resolved or, equivalently, something about the highest spatial frequency that can be reproduced.

The smallest elements that can be created by a digital subsystem are the scenel, pixel, datel, and disel. These are created by a scene simulator, detector array, ADC, and the display mediums, respectively. The resel is the smallest element that can be created by an analog system. The optics, analog electronics, and video standard (or transmission link) each have their own resel. *System resolution is related to the largest "-el."*

Often, the display is the limiting factor in terms of image quality and resolution. No matter how good the electronic imaging system is, if the display resolution is poor, then the overall system resolution is poor. But if the system is optics-limited, then a high resolution display does not offer any "extra" system resolution. A high resolution display just ensures that all the information available is displayed. System resolution may be limited by the HVS. If the observer is too far from the screen, not all of the image detail can be discerned.

Complex systems cannot be characterized by a single number (for example, an "-el"). Nevertheless, this chapter discusses various "-els" because they have been in existence so long. They were developed to describe the resolution of specific subsystems; "-els" do not provide an end-to-end performance metric. Nor do they include aliasing. These important considerations are discussed in Chapter 11, *Image Quality Metrics*.

The symbols used in this book are summarized in the *Symbol List* (page xiii) which appears after the *Table of Contents*.

10.1. ELECTRONIC IMAGING SYSTEM RESOLUTION

Resolution provides valuable information regarding the finest spatial detail that can be discerned. A large variety of resolution measures exist[1] and the various definitions may not be interchangeable. An electronic imaging system is composed of many subsystems and each has its own metric for resolution (Table 10-1).

The attractiveness of resolution is that the maximum range at which a target can be detected is easily estimated. A back-of-the-envelope approximation provides the range:

$$Range = \frac{target\ size}{resolution}. \qquad (10\text{-}1)$$

where "resolution" has been expressed in object space units. Equivalently, when the distance is known, then the resolution specifies the smallest target that can be resolved.

Table 10-1
MEASURES OF RESOLUTION

SUBSYSTEM	RESOLUTION METRIC
Optics	Rayleigh criterion Airy disk diameter Blur diameter
Detectors	Detector-angular-subtense Instantaneous-field-of-view Effective-instantaneous-field-of-view Detector pitch Pixel-angular-subtense
Electronics	Bandwidth
Electronic imaging system (MTF approach)	Limiting resolution Nyquist frequency
Displays	TV limiting resolution

276 SAMPLING, ALIASING, and DATA FIDELITY

10.1.1. OPTICAL RESOLUTION METRICS

Table 10-2 provides various optical resolutions. As analog metrics, these are various definitions of a resel. Diffraction measures include the Rayleigh criterion and the Airy disk diameter. The Airy disk is the bright center of the diffraction pattern produced by an ideal optical system. In the focal plane of the lens, the Airy disk diameter is

$$d_{AIRY} = 2.44 \frac{\lambda}{D_o} fl = 2.44 \lambda F . \qquad (10\text{-}2)$$

The Rayleigh criterion is a measure of the ability to distinguish two closely spaced objects when the objects are point sources. Optical aberrations and focus limitations increase the diffraction-limited spot diameter to the blur diameter. Optical designers use ray tracing programs to calculate the blur diameter. The blur diameter size is dependent on how it is specified (i.e., the fraction of encircled energy[2]). den Dekker and van den Bos provide[3] an in-depth review of optical resolution metrics.

Considerable literature[4] has been written on the image forming capability of lens systems. Image quality metrics include aberrations, Strehl ratio, and blur or spot diagrams. These metrics, in one form or another, compare the actual blur diagram to diffraction-limited spot size. As the blur diameter increases, image quality decreases and edges become fuzzy. The ability to see this degradation requires a high resolution sensor. Both the human eye and photographic film visually have this resolution. However, with electronic imaging systems, the detectors are often too large to see the degradation because the detectors are often larger than the blur diameter.

Table 10-2
OPTICAL RESOLUTION METRICS (In image space)

RESOLUTION	DESCRIPTION	DEFINITION (usual units)
Rayleigh criterion	Ability to distinguish two adjacent point sources	$x_{RES} = 1.22 \lambda F$ (Calculated: mm)
Airy disk	Diffraction-limited diameter produced by a point source	$x_{RES} = 2.44 \lambda F$ (Calculated: mm)
Blur diameter	Actual minimum diameter produced by a point source	Calculated from ray tracing (mm)

10.1.2. DETECTOR RESOLUTION

Considering the Airy disk, the wavelength dependence suggests that the resolution increases as the wavelength decreases (smaller is better). However, this is only true when the detector size is much smaller than the Airy disk size. For most electronic imaging systems the reverse is true. The detector is the same size or larger than the disk diameter. This means that the Airy disk is not adequately sampled and therefore the resolution afforded by the optics cannot be exploited. Systems designed for star detection are usually optics-limited (resel larger than a pixel) whereas for general imagery, the systems are usually detector-limited (pixel larger than the resel).

Detector arrays are often specified by the number of pixels and detector pitch. These are not meaningful until an optical system is placed in front of the array. Table 10-3 provides the most common resolution metrics expressed in object space units. The detector-angular-subtense is often used by the military to describe the resolution of systems when the detector is the limiting subsystem. If the detector's element horizontal and vertical dimensions are different, then the DAS in the two directions is different. Note that the pixel-angular-subtense (PAS) is different than the detector-angular-subtense (DAS). Only with 100% fill-factor arrays are they the same. Spatial sampling rates are determined by the pitch in image space or the PAS in object space.

Table 10-3
DETECTOR ELEMENT RESOLUTION MEASURES (in object space)

RESOLUTION	DESCRIPTION	DEFINITION (usual units)
Detector-angular-subtense	Angle subtended by one detector element	d/fl (Calculated: mrad)
Instantaneous-field-of-view	Angular region over which the detector senses radiation	Measured width at 50% amplitude (mrad)
Pixel-angular-subtense	Angle subtended by one pixel	d_{cc}/fl (Calculated: mrad)
Effective-instantaneous-field-of-view	One-half of the reciprocal of the object-space spatial frequency at which the MTF is 0.5	Measured or calculated (mrad)
Ground sampled distance	Projection of detector pitch onto the ground	$d_{cc}R/fl$ (calculated: inches)
Nyquist frequency	One-half of the angle subtended by detector pitch	$d_{cc}/2fl$ (calculated: mrad)

If resolution is defined by the DAS, then a vanishingly small detector is desired. Simultaneously, the f-number must approach zero so the system is not optics-limited (see Figure 9-17, page 252). Although any detector size, optical aperture, and focal length can be chosen to select the resolution limit, the same parameters affect the sensitivity (see Equation 9-43, page 254). With practical optical systems, the minimum f-number is about 1.0 with a theoretical limit of 0.5.

The instantaneous-field-of-view (IFOV) is the linear angle from which a single detector element senses radiation. It is a summary measure that includes both the optical and detector responses. The IFOV is typically a measured quantity. Here, a point source transverses the detector element and the detector output is graphed as a function of angle. The IFOV is the full-width, one-half maximum amplitude of the resultant signal. If the optical blur diameter is small compared to the DAS, then the IFOV is approximately equal to the DAS.

Nyquist frequency limits the highest spatial frequency that can be faithfully reproduced. The detector pitch provides a limit of

$$u_N = \frac{d_{CC}}{2fl} . \qquad (10\text{-}3)$$

For 100% fill-factor arrays, $d = d_{CC}$.

The detector cutoff is 1/DAS in object space. With detector arrays, signal fidelity is limited to the Nyquist frequency. Here, the effective-instantaneous-field-of-view (EIFOV) offers an alternate measure of resolution (Figure 10-1). u_{EIFOV} is the spatial frequency at which the MTF is 0.5. Figure 10-2 illustrates EIFOV for a system dominated by $MTF_{OPTICS} MTF_{DETECTOR}$. For detector-limited systems (large $d/F\lambda$), the IFOV is approximately equal to the DAS. Calling 1/DAS the cutoff provides a misleading representation. Recall that the detector can respond to frequencies above 1/DAS (see Figure 5-5, page 105).

When viewing aerial imagery that contains a test pattern (such as the USAF 1951 standard three-bar target), an image analyst determines the smallest discernible cycle on the ground. This cycle width (bar plus space) is the ground resolved distance (GRD). It includes the system MTF and possible degradation by the atmosphere and is a function of altitude. The GRD is related[5] to the 10-point Image Interpretability Rating Scale (IIRS). The National Image Interpretability Rating Scale (NIIRS) has replaced the IIRS.

Until recently aerial imagery was captured exclusively on film. With the advent of electronic imaging systems, a related, but not identical, metric was created: the ground sampled distance (GSD). The GSD is the projection of the detector pitch onto the ground. It depends on the range from the system to the target and therefore is not unique. However, intelligence gathering tends to be collected from fixed altitudes (i.e., satellites and high altitude aircraft). The relationship between GRD (via NIIRS) and GSD has been empirically derived and it is called the general image quality equation (discussed in Section 11.10., *NIIRS*).

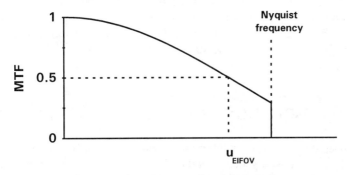

Figure 10-1. Definition of effective-instantaneous-field-of-view for an undersampled system. EIFOV = $1/(2\,u_{EIFOV})$. The cutoff at the Nyquist frequency is for convenience. It only represents the fact the signals above u_N cannot be faithfully reproduced. This misleading representation implies that there is *no* response above u_N. Frequencies above u_N are aliased.

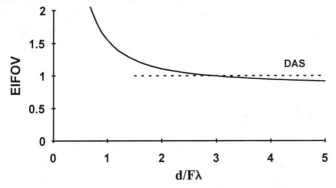

Figure 10-2. EIFOV as a function of $d/F\lambda$ for square detectors ($d = d_H = d_V$). The DAS has been normalized to unity.

280 SAMPLING, ALIASING, and DATA FIDELITY

10.1.3. ELECTRICAL RESOLUTION METRIC

For high data rate systems such as line-scanners, the electronic bandwidth may limit system response. For electronic circuits, resolution is implied by its bandwidth. The minimum pulse width is approximately

$$\tau_{MINIMUM} \sim \frac{1}{2\,BW}, \qquad (10\text{-}4)$$

where $\tau_{MINIMUM}$ is measured in seconds and BW is measured in hertz. The 3-dB point (half-power frequency) is often called the bandwidth. With this minimum value the output signal *looks* like the input. Other expressions (e.g., $\tau_{MINIMUM}$ = 1/BW) simply imply there is a measurable output without any inference about signal fidelity.

10.2. CRT RESOLUTION

CRT resolution metrics are intertwined with television standards such that they are often presented together. Nearly all television receivers are built to the same basic design so that by specifying EIA 170 or NTSC, the television receiver bandwidth and spot size is implied. In this sense, a video standard has a "resolution."

Resolution is independent of display size. Displays can be built to any size and the viewing distance affects the *perceived* resolution. For example, if a display is designed for viewing at 45 inches, then it may have a 10-inch diagonal. If viewing will be at 90 inches, the entire display can simply be made larger (20-inch diagonal). The larger display has larger raster lines. But the video electronic bandwidth is the same and the resolution remains constant. When viewed at 45 inches, the larger display will appear to have poorer image quality.

10.2.1. VERTICAL RESOLUTION

The number of raster lines limits the vertical resolution. For example, NTSC-compatible monitors display 485 lines. However, to see precisely this number, the test pattern must be perfectly aligned with the raster pattern. Because the test pattern may be randomly placed, this pattern can produce zero output if 180° out-of-phase (see Section 8.1., *Phasing Effects*, page 197). To account for random phases, the Kell factor[6] is applied to the vertical resolution to provide an average value:

$$R_{VERTICAL} \approx (active\,scan\,lines)(Kell\,factor) . \quad (10\text{-}5)$$

A value of 0.7 is widely used. Timing considerations force the vertical resolution to be proportional to the number of vertical lines in the video signal. Therefore, NTSC displays have an average vertical resolution of (0.7)(485) = 340 lines. PAL and SECAM displays offer an average of (0.7)(575) = 402 lines of resolution.

10.2.2. THEORETICAL HORIZONTAL RESOLUTION

The theoretical horizontal resolution is usually normalized to the picture height. For a video bandwidth of BW, the theoretical resolution is

$$R_{TVL} = \frac{(2BW)(active\,line\,time)}{aspect\,ratio} \frac{TVL}{PH} . \quad (10\text{-}6)$$

The units are television lines per picture height (TVL/PH). Note that there are two TV lines per cycle. While the video bandwidth can be any value, for consumer applications, it matches the video standard. For NTSC television receivers, the nominal horizontal resolution is (2)(4.2 MHz)(52.45 μs)/(4/3) = 330 TVL/PH. PAL and SECAM values are approximately 426 and 465 TVL/PH, respectively. These values change slightly depending on the active line time and bandwidth selected. The theoretical horizontal resolution is approximately equal to the vertical resolution.

Displays designed for scientific applications usually require the standard video timing format but have a much wider video bandwidth. These monitors can display more horizontal TV lines per picture height than that suggested by, say, the NTSC video transmission bandwidth. The horizontal bandwidth can be any value and therefore the theoretical horizontal resolution can be any value.

10.2.3. TV LIMITING RESOLUTION

TV limiting resolution is a measure of when alternate vertical bars are just visible. The standard resolution test target is a wedge pattern with spatial frequency increasing toward the apex of the wedge. It is equivalent to a variable square wave pattern.

The measurement is a perceptual one and the results vary across the observer population. The flat field condition and high TV limiting resolution are conflicting requirements. For high TV limiting (horizontal) resolution, σ_{SPOT} must be small. But raster pattern visibility (more precisely, invisibility) suggests that σ_{SPOT} should be large. The ratio of spot size to raster pitch is the resolution/addressability ratio[7] (RAR). For most displays, RAR $\approx$ 1. Then the TV resolution is approximately equal to the theoretical horizontal resolution.

TV limiting resolution is subjective and occurs when the MTF is approximately 3%. For commercial applications, the resolution is often specified by the spatial frequency where the MTF is 10%.

10.3. MTF-BASED RESOLUTION

Usually systems are designed to have high MTF and Figure 9-17 (page 252) illustrated the MTF at Nyquist frequency. However, the value of Nyquist frequency is a measure of resolution; 100% fill-factor staring arrays have the highest Nyquist frequency.

10.3.1. LIMITING RESOLUTION

Other measures also exist such as the limiting resolution. This is the spatial frequency at which the MTF is, say, 5%. Figure 10-3 illustrates two systems that have the same limiting resolution. Which system is selected depends on the specific application. System A is better at high spatial frequencies, and system B is better at low spatial frequencies. Equivalently, if the scene contains mostly low frequencies (large objects), then system B should be used. If edge detection is important (high spatial frequencies), then system A should be considered. This clearly indicates the difficulty encountered when using resolution exclusively as a measure of system performance.

10.3.2. MTF/TM RESOLUTION

Early resolution measures were based on wet-film camera characteristics. The intersection of the system MTF and film threshold modulation (TM) provided a measure of camera *system* performance (Figure 10-4). The value u_{RES} is the limiting spatial frequency or intersection frequency and TM is the minimum modulation required at a given spatial frequency to discern a difference.

SYSTEM RESOLUTION 283

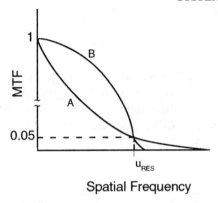

Figure 10-3. The MTFs of two different imaging systems with the same limiting resolution, u_{RES}.

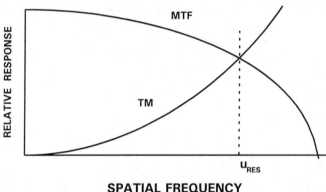

Figure 10-4. The spatial frequency of the intersection of film's threshold modulation and MTF_{SYSTEM} is a measure of resolution.

Conceptually, for spatial frequencies above u_{RES}, the film requires more modulation than that provided by the system. Below u_{RES} the film can easily discern the scene modulation. Because many measures were in support of aerial imagery, the intersection was also called the aerial image modulation (AIM). With the advent of electronic imaging systems, the film TM was replaced with the eye's threshold modulation[8] (discussed in Section 11.4., *MTFA*).

10.3.3. SÉQUIN'S LIMITING RESOLUTION

Different formulations are used to describe CRT vertical resolution. Séquin[9] defined resolution as the spatial frequency where the aliased component amplitude is one-half of the direct response. Assuming that MTF_{OPTICS} is very high, Figure 10-5 provides $MTF_{DETECTOR}$ and MTF_{CRT} for a 100% fill-factor array. For this example, $v_{RES} = v_{iD}/3$.

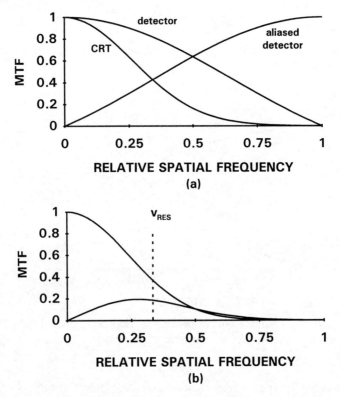

Figure 10-5. Séquin's resolution. (a) $MTF_{DETECTOR}$ and MTF_{CRT} for a 100% fill-factor array as a function of v_i/v_{iD}. (b) Direct and aliased components. $v_N = 0.5$ and $v_{RES} = 1/3$.

10.4. SHADE'S EQUIVALENT RESOLUTION

As reported by Lloyd[10], Sendall modified Shade's equivalent resolution[11] such that

$$R_{EQ} = \frac{1}{2N_e} = \frac{1}{2\int_0^\infty [MTF_{SYS}(u)]^2 \, du} , \qquad (10\text{-}7)$$

where N_e is Shade's equivalent pass band. R_{EQ} cannot be directly measured and is a mathematical construct used simply to express overall performance. As R_{EQ} decreases, the resolution "improves" (smaller is better). As an approximation, the system resolution, $R_{EQ\text{-}SYS}$, may be estimated from the component equivalent resolutions, R_i, by

$$R_{EQ\text{-}SYS} \approx \sqrt{R_1^2 + R_2^2 + \ldots + R_n^2} . \qquad (10\text{-}8)$$

Shade's approach using the square of the MTF emphasized those spatial frequencies at which the MTF is relatively high. It appears to be a good measure for classical systems in which the MTF is decreasing (such as a Gaussian distribution). The equivalent resolution approach assumes that the system is completely analog and it ignores sampling effects. Therefore, R_{EQ} becomes a resel. It probably should not be used (nor should any other image quality metric) to compare systems with significantly different MTFs. Nor should it be used with systems that can create significant aliasing.

As a summary metric, $R_{EQ\text{-}SYS}$ provides a better indication of system performance than just a single metric such as the DAS. $R_{EQ\text{-}SYS}$ probably should be used in Equation 10-1 (page 275) to obtain a more realistic measure of range performance. R_{EQ} cannot be evaluated in closed form for all subsystems such as boost, defocused optics or Chebyshev filters. For these subsystems, Equation 10-7 must be evaluated numerically. If a subsystem MTF is essentially unity over the spatial frequency region of interest, then that subsystem R_{EQ} can be ignored (e.g., $R_{EQ} = 0$). Table 10-4 provides the equivalent resolution for several common MTFs.

Table 10-4
ONE-DIMENSIONAL EQUIVALENT RESOLUTIONS

SUBSYSTEM	MTF	R_{EQ}
Optics	$\dfrac{2}{\pi}\left(\cos^{-1}\left(\dfrac{u_i}{u_{iC}}\right) - \dfrac{u_i}{u_{iC}}\sqrt{1-\left(\dfrac{u_i}{u_{iC}}\right)^2}\right)$	$1.845\,\lambda F$
Rectangular detector	$\text{sinc}(d\,u_i)$	d
N^{th}-order low-pass filter	$\dfrac{1}{\sqrt{1+\left(\dfrac{f_e}{f_{e3dB}}\right)^{2N}}}$	$\dfrac{1}{\pi f_{e3dB}}$ when $N=1$ $\dfrac{1}{2 f_{e3dB}}$ when $N \to \infty$
CRT-based display	$e^{-2\pi^2 \sigma_{SPOT}^2 u_d^2}$	$2\sqrt{\pi}\,\sigma_{SPOT}$

If the electronics provides adequate bandwidth, it is instructive to see the relationship between the Airy disk diameter and the detector size. Using the same parameters as Figure 9-17 (page 252), $x = d/(F\lambda)$. For a system whose resolution is limited by the optics/detector combination,

$$R_{EQ\text{-}SYS} = d\sqrt{\left(\frac{1.845}{x}\right)^2 + 1} \ . \tag{10-9}$$

SYSTEM RESOLUTION 287

The equation is graphed in Figure 10-6. As x increases, R_{EQ-SYS} approaches d. For small values of x, the system becomes optics-limited and the equivalent resolution increases. When x = 2.44, R_{EQ-SYS} = 1.25d and when x = 4, R_{EQ-SYS} = 1.10d. Matching the Airy disk to the detector size creates an equivalent resolution 25% larger than the actual detector size.

Both Figures 10-6 and 10-2 have the same functional form. Shade's equivalent resolution provides values about 20% larger than that predicted by the EIFOV. If, instead, the EIFOV was defined when the MTF = 0.6, then the predicted EIFOV will be approximately equal to Shade's equivalent resolution.

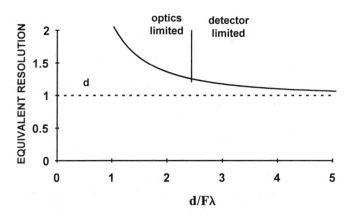

Figure 10-6. Equivalent resolution normalized to d as a function of d/Fλ. The vertical line indicates where the Airy disk diameter is equal to the detector size. The best resolution occurs when x approaches infinity. Here the optics MTF approaches unity.

10.5. SYSTEM RESOLUTION EXAMPLES

Example 10-1
PIXELS and DATELS

An imaging system consists of 512 × 512 detector elements. Assume that the fill-factor is 100% and the camera's analog output is digitized at either 1024 samples/line or 2048 samples/line. What is the resolution?

288 SAMPLING, ALIASING, and DATA FIDELITY

The external analog-to-digital converter creates either two or four datels for each pixel. In both cases, the pixel size remains the same. Increasing the sampling rate does not change the system's resolution but improves the repeatability of all measurements by reducing phasing effects that might occur in the analog-to-digital converter. Any image processing algorithm that operates on this higher number must consider the sensor resolution.

The highest spatial frequency that can be reproduced is limited by the array Nyquist frequency. If MTF_{OPTICS} is high and the analog bandwidth is sufficiently wide, then the minimum pulse width corresponds to a pixel width.

Cameras with a standard video output may not have the resolution suggested by the video standard. For example, a camera containing an array of 320 × 240 pixels may have its output formatted into EIA 170 timing. This standard suggests[8] that it can support approximately 438 × 480 pixels (see Section 6.6., *Frame Grabbers*, page 154). The signal may be digitized by a frame grabber that creates 640 × 480 datels. The image size is still 320 × 240 pixels.

Example 10-2
DISELS and PIXELS

A staring array consists of 512 × 512 detectors. Using 4× electronic zoom it is presented on a digital monitor that displays 1024 × 1024 disels. What is the system resolution?

Each pixel is mapped onto four datels that are then mapped one-to-one onto disels. Here, the imaging system determines the system resolution, not the number of datels or disels. High quality monitors only ensure that the image quality is not degraded. Electronic zoom cannot increase resolution. But decimation or minifying may reduce resolution. Because image processing is performed on datels, the image analyst must be made aware of the system resolution.

SYSTEM RESOLUTION 289

Example 10-3
SYSTEM CUTOFF

A staring array consists of detectors that are 9 μm × 9 μm in size. The detector pitch is 15 μm. The focal length is 15 cm. The aperture diameter is 3 cm and the average wavelength is 0.5 μm. What is the system cutoff?

System cutoff is defined as the smaller of the optical cutoff, detector cutoff, or Nyquist frequency. The optical cutoff is absolute. No input modulation with spatial frequency above the optical cutoff can be reproduced. The detector cutoff has been arbitrarily defined as 1/DAS. While frequencies above detector cutoff are faithfully reproduced, they suffer phase reversal (see Figure 5-5, page 105). With staring arrays, these frequencies are aliased. Selecting the Nyquist frequency as cutoff is only appropriate when the ideal anti-alias and reconstruction filters are used. Otherwise, systems respond to frequencies above Nyquist. The output is, of course, distorted. Selecting the Nyquist frequency as cutoff often provides a misleading representation of system performance. This is not in conflict with relating system resolution to the Nyquist frequency. It is the term *cutoff* that must be carefully interpreted.

In image space, the optical cutoff is $u_{iC} = D_o/(\lambda fl) = 400$ cycles/mm. The detector cutoff is $1/d = 111$ cycles/mm. The detector pitch provides sampling every 15 μm for an effective sampling rate at 66.7 cycles/mm with the Nyquist frequency being one-half that value. Because the Nyquist frequency is the smallest value, the "system cutoff" is 33.3 cycles/mm.

Example 10-4
EQUIVALENT RESOLUTION

An infrared imaging system operating in the long-wave infrared spectral region has an entrance diameter of 10 inches. The DAS is 0.1 mrad. The system contains a single-pole, low-pass filter whose f_{3dB} is 6 cycles/mrad when transposed to object space. What is the resolution?

290 SAMPLING, ALIASING, and DATA FIDELITY

Assuming $\lambda_{AVE} \approx 10\,\mu m$, the optics provide

$$R_{EQ} = 1.845\frac{\lambda}{D_o} = 1.845\frac{10\times10^{-6}\,m}{254\times10^{-3}\,m} = 0.0726\ mrad\ . \quad (10\text{-}10)$$

The detector provides $R_{EQ} = 0.1$ mrad. Here, $d/(F\lambda) = \alpha D_o/\lambda = 2.54$. The electronics provides

$$R_{EQ} = \frac{1}{\pi f_{3dB}} = \frac{1}{6\pi} = 0.0531\ mrad\ . \quad (10\text{-}11)$$

The resultant system resolution is

$$R_{EQ\text{-}SYS} = \sqrt{(0.0762)^2 + (0.1)^2 + (0.0531)^2} = 0.136\ mrad\ . \quad (10\text{-}12)$$

Example 10-5
SYSTEM EQUIVALENT RESOLUTION

A CCD camera has an analog output digitized by a frame grabber. What is the apparent system resolution for image processing and what is the resolution displayed on a CRT-based monitor?

This CCD camera contains 480 horizontal detectors. Each detector is 15 μm on 15 μm centers (100% fill-factor). The optical aperture is 1 inch and the focal length is 5.6 inches (F = 5.6). The analog electronics is wideband and it does not significantly affect the resolution. Camera output is in the standard RS 170 format with a bandwidth of 4.2 MHz and $t_{VIDEO\text{-}LINE} = 52.45\ \mu s$. The frame grabber provides 640 horizontal samples and a standard consumer CRT-based monitor is used.

Assuming $\lambda_{AVE} \approx 0.5\,\mu m$, R_{EQ} for the optics is $1.845\lambda_{AVE}/D_o = 0.0363$ mrad. The DAS is d/fl or 0.105 mrad. Here, $d/(F\lambda) = \alpha D_o/\lambda = 5.33$. That is, the optical/detector subsystem is detector-limited.

The HFOV is

$$HFOV = \frac{d_{ARRAY}}{fl} = \frac{480 \cdot 15 \times 10^{-6}}{5.6 \cdot 25.4 \times 10^{-3}} = 50.6\, mrad\ . \quad (10\text{-}13)$$

With an active line time of 52.45 μs, the video bandwidth in object space is

$$f_{o\text{-}VIDEO} = \frac{t_{VIDEO\text{-}LINE}}{HFOV} BW = \frac{52.45 \times 10^{-6}}{50.4 \times 10^{-3}} 4.2 \times 10^6 = 4.37\, \frac{cycles}{mrad}\ . \quad (10\text{-}14)$$

Assuming a sharp cutoff, $R_{EQ} \approx 1/2f_{3dB} = 0.114$ mrad. The system equivalent resolution is

$$R_{EQ\text{-}SYS} = \sqrt{0.0363^2 + 0.105^2 + 0.114^2} = 0.159\, mrad\ . \quad (10\text{-}15)$$

Here, the RS 170 bandwidth significantly affects the resolution. Neglecting phasing effects, each frame grabber datel represents approximately (0.159)(480)/640 = 0.119 mrad. These are overlapping resolution elements where the center-to-center spacing is (0.105)(480)/640 = 0.0787 mrad.

The CRT equivalent resolution is

$$R_{EQ} = 2\sqrt{\pi}\, \sigma_o = 2\sqrt{\pi}\left(\frac{d_{CC}}{1.645 \cdot fl}\right) = 0.227\, mrad\ . \quad (10\text{-}16)$$

The display significantly degrades the resolution. But this is necessary to suppress the raster and sampling artifacts. Each disel represents overlapping resolution elements whose apparent size is

$$R_{EQ\text{-}SYS} = \sqrt{0.159^2 + 0.227^2} = 0.277\, mrad\ . \quad (10\text{-}17)$$

The MTFs are plotted in Figure 10-7. Numerical integration of Equation 10-7 provides $R_{EQ\text{-}SYS} = 0.263$. For this example, Equations 10-7 and 10-8 provided similar results.

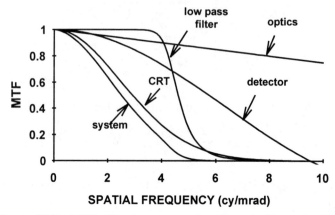

Figure 10-7. MTFs for the system described in Example 10-5. The video electronics is modeled as a Butterworth filter where N = 10.

10.6. REFERENCES

1. G. C. Holst, *Electro-Optical Imaging System Performance*, pp. 218-228, JCD Publishing, Winter Park, FL (1996).
2. L. M. Beyer, S. H. Cobb, and L. C. Clune, "Ensquared Power for Obscured Circular Pupils With Off-Center Imaging," *Applied Optics*, Vol. 30(25), pp. 3569-3574 (1991).
3. A. J. den Dekker and A. van den Bos, "Resolution: A Survey," *Journal of the Optical Society A*, Vol. 14(3), pp. 547-557 (1997).
4. See, for example, W. J. Smith, *Modern Optical Engineering*, 2nd Edition, McGraw-Hill, New York (1990).
5. Air Standardization Agreement: "Imagery Interpretability Rating Scale," Air Standardization Co-ordinating Committee report AIR STD 101/11 dated 10 July 1978.
6. S. C. Hsu, "The Kell Factor: Past and Present," *SMPTE Journal*, Vol. 95, pp. 206-214 (1986).
7. G. C. Holst, *CCD Arrays, Cameras, and Displays*, pp. 189-194, JCD Publishing, Winter Park, FL (1996).
8. H. L. Synder, "Image Quality and Observer Performance," in *Perception of Displayed Information*, L. M. Biberman, ed., pp. 87-118, Plenum Press, New York, NY (1973).
9. C. H. Séquin, "Interlacing in Charge-coupled Imaging Devices," *IEEE Transactions on Electron Devices*, Vol. ED-20, pp. 535-541 (1973).
10. J. M. Lloyd, *Thermal Imaging*, page 109, Plenum Press, New York (1975).
11. O. H. Shade, Sr., "Image Gradation, Graininess, and Sharpness in Television and Motion Picture Systems," published in four parts in *SMPTE Journal*: "Part I: Image Structure and Transfer Characteristics," Vol. 56(2), pp. 137-171 (1951); "Part II: The Grain Structure of Motion Pictures - An Analysis of Deviations and Fluctuations of the Sample Number," Vol. 58(2), pp. 181-222 (1952); "Part III: The Grain Structure of Television Images," Vol. 61(2), pp. 97-164 (1953); "Part IV: Image Analysis in Photographic and Television Systems," Vol. 64(11), pp. 593-617 (1955).

11

IMAGE QUALITY METRICS

Our perception of good image quality is based on the real-world experiences of seeing all colors, intensities, and textures. An imaging system has a limited field-of-view, limited temporal and spatial resolutions, and presents a two-dimensional view of a three-dimensional world. In the real-world our eyes scan the entire scene. An imaging system cannot capture all of the available information. Furthermore, an imaging system introduces noise and the loss of image quality due to noise can only be estimated.

Image quality is a subjective impression ranging from poor to excellent. It is a somewhat learned ability. It is a perceptual one, accomplished by the brain, affected by and incorporating inputs from other sensory systems, emotions, learning and memory. The relationships are many and still are not well understood. Perceptual quality of the same scene varies between individuals and temporally for the same individual. Large variations exist in an observer's judgment as to the correct rank ordering of image quality from poor to best. Image quality cannot be placed on an absolute scale.

Many formulas exist for predicting image quality. Each is appropriate under a particular set of viewing conditions. These expressions are typically obtained from empirical data in which multiple observers view many images with a known amount of degradation. The observers rank the imagery from worst to best and then an equation is derived which relates the ranking scale to the amount of degradation.

All current image quality metrics are only a slice through a multidimensional space that appears to include nearly an infinite number of parameters (e.g., MTF, noise, contrast, frame rate, etc.). The slice may only consider one component (e.g., image quality as a function of MTF). The resolution measures discussed in Chapter 10, *System Resolution*, provided a point in this multidimensional space.

Early metrics were created for film-based cameras. Image quality was related to the camera lens and film MTFs. With the advent of television, image quality centered on the perception of raster lines (sampling effects) and the minimum SNR required for *good* imagery.

Biberman[1] asks *What does sampling do to picture quality?* and *What must the sampling frequency and format be to minimize image deterioration?* He further states: *In the case of image formation by a matrix of detectors with electrical outputs, the problems associated with fabrication, cost, interconnections of wiring, number of amplifiers, and so forth serve to constrain the number of detectors used. Thus there tend to be many fewer detector elements per picture...and the limitations associated with sampling the imagery must be considered.... Basically, there are three factors affecting this problem of finite sampling and image quality: (1) The number of samples per image, (2) The signal-to-noise ratio per sample. (3) The generation of spurious signals by the sampling process.*

With the desire to design the least-complex system, Biberman's questions lead to *How much aliasing can be tolerated?* It would appear that aliasing is a serious problem. But, the naturally occurring real world is typically aperiodic and the moiré structure is rarely seen in real imagery. Legault[2] looked at 109 images before finding an example of obvious aliasing. He selected a plowed field. Whether aliasing is bothersome (it is always present) is scene dependent and the degree of undesirability cannot be predicted in advance.

We have become accustomed to undersampling. Commercial television is undersampled in the vertical direction due to the raster pattern and, as CCD cameras replace vidicons, TV is undersampled in both directions. The undersampling effect produces moiré patterns that become evident when the image contains periodic objects - such as a person wearing a striped shirt. We have become accustomed to this "standard TV quality" and an image must be significantly degraded before we object to the image quality. However, this does not mean that aliasing should be neglected. It introduces artifacts in all imagery.

It would appear that aliasing is an extremely serious problem. However, the extent of the problem depends on the final interpreter of the output data. In general, electronic signals are band-limited or can be limited by a low-pass filter (anti-alias filter). Data fidelity then centers on the ability to reproduce a step or pulse. But electronic imaging systems inherently undersample the scene and alias the signal. More literature exists on aliased imagery and image quality metrics than electronic signal reproduction.

A comprehensive end-to-end analysis should include both aliased signal and aliased noise. This model starts with the scene characteristics and ends with the human interpretation of the displayed information. Park and Hazra[3] consider aliased signal to be part of the noise spectrum because it interferes with the ability to perceive targets. As a signal dependent value, it can only be analyzed

IMAGE QUALITY METRICS 295

on a case-by-case basis. An MTF and associated aliasing may be acceptable for some imagery and not for others. As such, no single image quality metric can apply to all images. Boost enhances frequency amplitudes where the MTF is low. This tends to be the same region that contains aliased signal and aliased noise. These aliased components limit[4] the extent to which a sampled image can be sharpened (see Figure 9-10, page 244).

An aliased signal, while mathematically describable, affects image quality in an unknown fashion. Image quality metrics that consider aliased signals are only approximations. Although no universal method exists for quantifying aliasing, it is reasonable to assume that it is proportional to the MTF that exists above the Nyquist frequencies: u_N and v_N. Whether this aliasing is objectionable (it varies with individuals), depends on the electronic imaging system application. Any image quality metric must be used with caution when significant aliasing may occur.

There are potentially two different system design requirements: (1) Good image quality and (2) performing a specific task. Sometimes these are equivalent, but other times they are not. While good image quality is always desired, a military system designed to detect a specific target may not provide the "best" image quality. Computer monitors are usually designed to make alphanumeric characters readable (specific application). They also provide good imagery.

When the final image is compared to the original, image quality focuses on image fidelity: The ability to reproduce the original image precisely. However, with television and recorded video, the original scene is not available for viewing. Here, image quality depends on our experience. "Good" images may have been modified by image enhancement algorithms.

A large number of metrics are related to image quality. Most are based on monochrome imagery such as resolution, MTF, and minimum resolvable contrast. They are concerned with human interpretation of image quality. The metrics presented here do not apply to the recognition of alphanumeric characters. This has been extensively studied by those interested in optimizing computer displays. Alphanumeric character recognition metrics are based primarily on the number of pixels per character. Color reproduction[5] and tonal transfer issues,[6] very important to color cameras, are not covered here.

The symbols used in this book are summarized in the *Symbol List* (page xiii) which appears after the *Table of Contents*.

296 SAMPLING, ALIASING, and DATA FIDELITY

11.1. IMAGE QUALITY MODEL

The c/d/c/d/c model presented in Figure 1-17 (page 24) provides the frame work for discussing image quality. Figure 11-1 illustrates the various inputs and outputs that can be evaluated. The value f_i is the input to subsystem i and g_i is its output. The output of the previous subsystem, g_{i-1}, is the input to the current subsystem. That is, $f_i = g_{i-1}$. When specified in the frequency domain, the ratio becomes that subsystem's MTF

$$MTF_i(u,v) = \frac{G_i(u,v)}{F_i(u,v)} \ . \tag{11-1}$$

For brevity, the difference between $f_i(x,y)$ and $g_j(x,y)$ will be denoted as $[f_i,g_j]$ and the ratio of $G_i(u,v)$ and $F_j(u,v)$ as $[F_i,G_j]$.

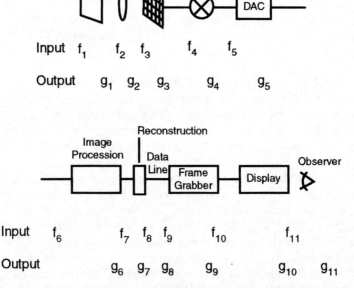

Figure 11-1. Many image quality relationships apply only to part of the electronic imaging system. f_i is the input to subsystem i and g_i is the output. Although noise can be added to all subsystems, the detector creates the dominant noise component. (After Reference 7).

IMAGE QUALITY METRICS 297

Many resolution and image quality metrics are based on $[f_i, g_j]$, $[f^2_i, g^2_j]$, $[F_i, G_j]$, $[F^2_i, G^2_j]$, etc. For example, optical design measures such as the Rayleigh criterion, blur diameter, and Strehl ratio were derived from $[f_2, g_2]$ or $[F_2, G_2]$. Aliasing can occur at $[f_3, g_3]$, $[f_5, g_5]$, $[f_9, g_9]$, and $[f_{10}, g_{10}]$. The end-to-end system performance is related to $[f_1, g_{10}]$ without the observer and $[f_1, g_{11}]$ with the observer. The function $F_2(u,v)$ describes the scene spatial content. This was previously labeled as the object, $O(u,v)$. Chapter 8, *Reconstructed Signal Appearance*, illustrated $[f_2, g_3]$ whereas MTF analysis employs $[F_i, G_j]$ to describe system performance.

Some image processing algorithms consider only $[f_6, g_6]$. Data links are described by physical measurables such as bandwidth and rise time. If the bandwidth is appropriate, then g_8 is, to within a specified tolerance, equal to f_8. Display manufacturers use $[f_{10}, g_{10}]$ as a design metric but the effectiveness of the design can only be interpreted by the observer.

As image quality increases, the output appears more like the input. Quality may be proportional to the difference at a discrete value, as an area difference, mean square error, or energy difference, respectively:

$$Q_{VALUE} = k\,[g_j(x_o, y_o) - f_i(x_o, y_o)] \;, \tag{11-2}$$

$$Q_{AREA} = k \iint [g_j(x,y) - f_i(x,y)]\, dx\, dy \;, \tag{11-3}$$

$$Q_{MSE} = k \iint |g_j(x,y) - f_i(x,y)|^2\, dx\, dy \;, \tag{11-4}$$

and

$$Q_{ENERGY} = k \iint [g_j^2(x,y) - f_i^2(x,y)]\, dx\, dy \;. \tag{11-5}$$

298 SAMPLING, ALIASING, and DATA FIDELITY

All these metrics can be "normalized." For example, the normalized mean square error is

$$Q_{NMSE} = k \frac{\iint |g_j(x,y) - f_i(x,y)|^2 \, dx\, dy}{\iint |f_i(x,y)|^2 \, dx\, dy} \, . \tag{11-6}$$

These could also be evaluated in the frequency domain

$$Q_{VALUE} = k \frac{G_j(u_o, v_o)}{F_i(u_o, v_o)} \, , \tag{11-7}$$

$$Q_{MSE} = k \iint |G_j(u,v) - F_i(u,v)|^2 \, du\, dv \, , \tag{11-8}$$

$$Q_{AREA} = k \frac{\iint G_j(u,v) \, du\, dv}{\iint F_i(u,v) \, du\, dv} \, , \tag{11-9}$$

and

$$Q_{ENERGY} = k \frac{\iint G_j^2(u,v) \, du\, dv}{\iint F_i^2(u,v) \, du\, dv} \, . \tag{11-10}$$

By including $F_2(u,v)$, the image quality metric becomes a function of the specific scene selected. For convenience, the scene is assumed to contain all frequencies over the region of interest: $F_2(u,v) = O(u,v) \approx 1$. For discrete subsystems (e.g., ../d/..), the integrals are replaced by summations. When noise is added, these expressions may become complex.

The myriad of image quality metrics suggests that no one metric can predict quality for all images. That is, the metrics do not always correlate well with visual perception. A specific metric may be adequate for a limited subset of images. Specific scenes can always be selected where minimal correlation exists between subjective impression and the objective quality metric. Conversely, an author can select specific images to "prove" correlation. Nevertheless, they do provide insight into the image formation and reconstruction processes. The mean square error is used most frequently.[7]

IMAGE QUALITY METRICS 299

Considering positive frequencies, the one dimensional displayed image is

$$F_{11}(u) = I_D(u) = MTF_{POST}(u) \sum_{n=0}^{\infty} MTF_{PRE}(nu_S \pm u) O(nu_S \pm u) , \quad (11\text{-}11)$$

where MTF_{PRE} contains all the MTFs up to the sampler (taken as the detector). It consists of the optical and detector MTFs. MTF_{POST} represents all the filters after the sampler. This includes the reconstruction filter and display. This equation can be rewritten as

$$I_D(u) = MTF_{POST}(u) MTF_{PRE}(u) O(u)$$

$$+ MTF_{POST}(u) \sum_{n=1}^{\infty} MTF_{PRE}(nu_S \pm u) O(nu_S \pm u) . \quad (11\text{-}12)$$

The first term is the spectrum of the displayed image where no sampling is present and is sometimes called the direct response. $MTF_{SYSTEM} = MTF_{PRE}MTF_{POST}$. The remaining terms represent aliasing. The ratio of aliased signal to total signal or the ratio of aliased energy to direct response energy can be used as a metric. In two dimensions,

$$Q_{ALIASED\;SIGNAL} = k \frac{\sum_m \sum_n \iint G_i(nu_s \pm u, mv_s \pm v) \, du\, dv}{\iint G_i(u,v) \, du\, dv} , \quad (11\text{-}13)$$

and

$$Q_{ALIASED\;ENERGY} = k \frac{\sum_m \sum_n \iint G_i^2(nu_s \pm u, mv_s \pm v) \, du\, dv}{\iint G_i^2(u,v) \, du\, dv} . \quad (11\text{-}14)$$

By using $G_i(u,v)$ or $G^2_i(u,v)$ in the numerator and denominator, the scene components are included. Thus, these metrics are scene dependent.

The ultimate aim is to create a metric that guides system design. It would be also convenient if this metric could be measured with test equipment to verify design and ensure quality control. Ideally, an objective measure will provide the same result as a subjective one. The metric may vary by application.

300 SAMPLING, ALIASING, and DATA FIDELITY

Image quality may be constrained by input/output devices, bandwidth limitations when transmitting data, storage constraints, or viewing conditions. Evaluation of only one subsystem response cannot optimize the overall system response. Efficient image processing cannot remove or even compensate for the aliasing that occurs at the detector. No complete model exists for the end-to-end performance, $[f_1,g_{11}]$. What follows are different metrics based on portions of $[f_i,g_j]$ or $[F_i,G_j]$.

11.2. MTF

Because of its frequency dependence, the MTF is more descriptive of system performance than a single value such as resolution. In general, images with higher MTFs and less noise are judged as having *better* image quality. There is no single *ideal* MTF shape that provides best image quality. According to Kusaka,[8] the MTF that produced the most aesthetically pleasing images depended on the scene content. Images with sharpened edges are usually judged to have better image quality. High frequencies can be restored with a boost filter. Because edges are associated with high frequencies, this may be considered a form of edge sharpening.

Most image quality metrics incorporate some form of the system MTF. The underlying assumption is that the image spectrum is limited by the system MTF. Equivalently, it is assumed that the scene contains all spatial frequencies and that the displayed image is limited by system MTF - a reasonable assumption for general imagery. While good image quality is always desired, a military system is designed to detect and recognize specific targets. Optimized military systems will have high MTF at the target frequencies and other spatial frequencies are considered less important.

The area under the MTF curve can also be used as a measure of image quality. Here, slight variations in the system spatial frequency cutoff generally do not affect this definition of image quality. Figure 11-2 illustrates two different systems whose area quality metric are the same. That is, the areas under the curves are equal. If the scene contains mostly low frequencies (large objects), then system B should be used. If edge detection is important (high spatial frequencies), then system A should be considered. When systems have different MTFs, specific scenes can be selected that make one system appear better than the other. This is particularly true if the MTFs are not similar in shape or functional form.

Most metrics using the MTF approach assume rotational symmetry. However, for many systems the vertical and horizontal MTFs are different. Two-dimensional image quality metrics (different vertical and horizontal MTFs) are formative at this time.

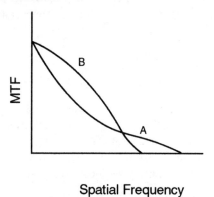

Figure 11-2. The MTFs of two different imaging systems. The areas under the curves are the same.

11.3. EQUIVALENT PASS BAND

Shade[9] claimed that the apparent image sharpness of a TV picture can be described by

$$N_e = \int_0^\infty [MTF(u)]^2 \, du , \quad (11\text{-}15)$$

where N_e is Shade's equivalent pass band. It is equal to the total energy that the system can transmit. By using the square of the MTF, it emphasizes those spatial frequencies at which the MTF is relatively high. It appears to be a good measure for classical systems in which the MTF is monotonically decreasing. The two MTFs shown in Figure 11-2 *could* have the same equivalent pass bands. Thus, a single number such as N_e should not be used to compare systems built to different designs. However, N_e is useful for comparing systems built to similar designs. N_e can be related to an equivalent resolution (see Section 10.4., *Shade's Equivalent Resolution*, page 285).

11.4. MTFA

The area bounded by the system MTF and the eye's threshold modulation (M_t) is called the modulation transfer function area (MTFA) (Figure 11-3a). M_t is also called the demand modulation function. In one-dimension,

$$MTFA = \int_0^{u_{RES}} \left[MTF_{SYSTEM}(u) - M_t(u) \right] du \ . \tag{11-16}$$

As the MTFA increases, the perceived image quality[10] appears to increase. The spatial frequency at which the MTF and M_t intersect is a measure of the overall resolution (see Section 10.3.2., *MTF/TM Resolution*, page 282). However, the resolution is not unique because M_t depends on the display viewing distance and lighting conditions. The eye's inhibitory response (see Figure 7-11, page 167) is often omitted from the MTFA approach.

According to Snyder,[10] the MTFA appears to correspond well with performance in military detection tasks where the targets are embedded in noise. The noise elevates the eye's threshold modulation so that the MTFA decreases with increasing noise (Figure 11-3b). For low-noise general imagery, the eye's contrast sensitivity has minimal effect on the MTFA. Then the integrand in Equation 11-16 can be replaced with just the system MTF.

11.5. SUBJECTIVE QUALITY FACTOR

Research results at Eastman Kodak suggested[11,12] that the spatial frequencies important to image quality are in the region from approximately one-third to three times the peak sensitivity of the HVS. This roughly covers the region between the 50% points on the HVS MTF. The HVS models of Nill,[13] Schultz,[14] and deJong and Bakker[15] support the Eastman Kodak approach. The Campbell-Robson data[16] can be approximated by

$$MTF_{EYE}(u_{eye}) = 10^{-1.4 \left[\log_{10} \left(\frac{u_{eye}}{u_{peak}} \right) \right]^2} \ . \tag{11-17}$$

The Kornfeld-Lawson model (Equation 9-39, page 251) does not include the inhibitory response and therefore cannot be used in the subjective quality factor (SQF).

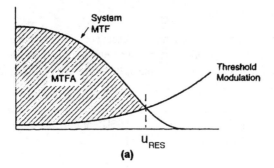

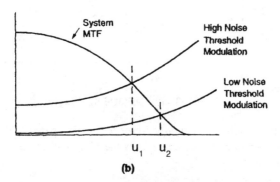

Figure 11-3. MTFA. The area between system MTF and threshold modulation is a measure of image quality. The spatial frequency at the intersection is a measure of limiting resolution. The area and limiting resolution depend on the noise level within the imagery. (a) Low noise and (b) elevated noise.

Assuming peak sensitivity at $u_{peak} = 4.5$ cycles/deg the SQF is:

$$SQF = \int_{\log(1.5)}^{\log(13.5)} MTF_{SYSTEM}(u_{eye}) \, d(\log(u_{eye})) \; . \qquad (11\text{-}18)$$

Figure 11-4 illustrates the SQF region. Because the eye response appears to be log-normally distributed, a quality factor based on a logarithmic scale appears reasonable. In the SQF approach, only those frequencies that are very important to the eye are included. The spatial frequency presented to the eye depends on the image size on the display, the distance to the display, and electronic zoom. As the SQF increases, image quality increases[12] (Table 11-1).

304 SAMPLING, ALIASING, and DATA FIDELITY

These results are based on many observers viewing noiseless photographs with known MTF degradation. If boost is employed, then it should be used to increase those spatial frequencies that fall within the SQF region.

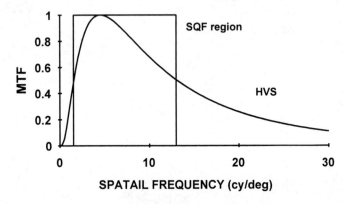

Figure 11-4. Subjective quality factor region compared to the Campbell-Robson HVS data. If boost increases the MTF within the rectangle, then the SQF increases.

Table 11-1
SUBJECTIVE QUALITY FACTOR

SQF	SUBJECTIVE IMAGE QUALITY
0.92	Excellent
0.80	Good
0.75	Acceptable
0.50	Unsatisfactory
0.25	Unusable

11.6. SQUARE-ROOT INTEGRAL

From an analyst's viewpoint, the convolution theorem states that MTFs should be multiplied, whereas the MTFA subtracted MTFs. Barten introduced the square-root integral (SQRI) approach[17] that overcame some of the limitations of the MTFA:

$$SQRI = \int_0^\infty \sqrt{MTF_{SYSTEM}(u) MTF_{HVS}(u)} \, d(\log(u)) \ . \quad (11\text{-}19)$$

This model includes the effects of various monitor parameters such as resolution, addressability, contrast, luminance, display size, and viewing distance. Barten expanded the model to include the eye's response to noise.[18,19]

Barten's approach is now more comprehensive in that it includes a variety of display parameters. It now includes[20] contrast, luminance, viewing ratio, number of scan lines, and noise. While Barten has incorporated the sampling effects of flat panel displays, no model includes the sampling (and associated aliasing) that takes place at the detector. Until this aliasing is quantified, no metric will predict image quality for sampled data systems. The equivalent pass band, SQF, and the SQRI provide similar results for systems with well-behaved MTFs. That is, higher values provide better imagery.

11.7. MRT and MRC

The perceived signal-to-noise ratio is

$$SNR_p = k \frac{MTF_{SYSTEM} \, \Delta I}{(system\ noise)} \frac{1}{(HVS\ spatial\ filter)(HVS\ temporal\ filter)} \ , \quad (11\text{-}20)$$

where ΔI is the intensity difference between the target and its immediate background and k is a proportionality constant that depends on the aperture diameter, focal length, quantum efficiency, etc. The HVS provides both temporal and spatial integration. It improves our ability to perceive targets embedded in noise. For modeling purposes, the eye-brain interpretative process acts as if spatial and temporal "filters" exist. These filters, which do not actually exist, reduce the perceived noise and thereby increase the perceived SNR.

When inverted, the minimum perceivable signal is proportional to

$$\Delta I = k' \frac{(system\ noise)}{MTF_{SYSTEM}} (HVS\ spatial\ filter)(HVS\ temporal\ filter)\ ,\quad (11\text{-}21)$$

where k' incorporates k and SNR_p. This equation is used by the military to calculate the minimum resolvable temperature[21] (MRT) and minimum resolvable contrast[22] (MRC). All calculations are performed in object-space spatial frequency with units of cycles/mrad.

MRT and MRC predict an observer's ability to detect a target embedded in noise. It combines resolution (as specified by the MTF) and sensitivity (as specified by the noise) into a composite figure of merit. Figure 11-5 illustrates the typical tradeoff between resolution and sensitivity. MRT and MRC do not include aliased signal and they limit predictions to the Nyquist frequency (Figure 11-6). This may have an excessive impact on range predictions.[23]

The probability of detection, recognition, and identification can be related to the number of cycles, N_{CYCLE}, across a target's critical dimension, $D_{CRITICAL}$. The critical dimension is the geometrical mean of the height and width of a rectangular target. The 50% probability is labeled as N_{50} (Table 11-2).

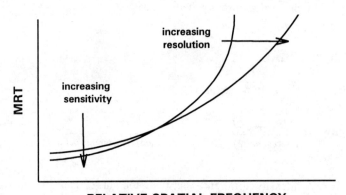

Figure 11-5. Tradeoff between sensitivity and resolution. For many systems, as resolution increases, sensitivity decreases. The Nyquist frequency limitation is not shown. The MRT is plotted on a logarithmic scale.

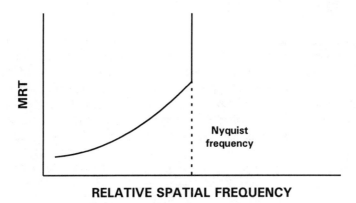
Figure 11-6. MRT artificially limited by Nyquist frequency.

Table 11-2
NUMBER OF CYCLES
(50% probability)

TASK	N_{50}
Detection	0.75
Recognition	3.0
Identification	6.0

The probability of other confidence levels is given by

$$P(N_{CYCLE}) = \frac{\left(\dfrac{N_{CYCLE}}{N_{50}}\right)^{2.7+0.7(N_{CYCLE}/N_{50})}}{1+\left(\dfrac{N_{CYCLE}}{N_{50}}\right)^{2.7+0.7(N_{CYCLE}/N_{50})}}, \quad (11\text{-}22)$$

where P(N_{CYCLE}) ranges from zero to one (0 to 100%). With a target discrimination metric (e.g., number of cycles across the critical dimension), the MRT or MRC spatial frequency axis is transformed into range:

$$u_o = N_{CYCLE}\frac{RANGE}{D_{CRITICAL}}\quad\frac{cycles}{mrad}. \quad (11\text{-}23)$$

11.8. ALIASED SIGNAL

Implicit in the sampling theorem is that the MTF is zero for all frequencies above the Nyquist frequency. The only problem here is that a system that has an MTF of zero for frequencies greater than u_N is likely to have an MTF much less than unity for frequencies approaching u_N. From a sampling theory point of view, the system MTF should be rectangular up to u_N. But this ideal shape is physically unrealizable. Furthermore, a system with a sharp transition in the frequency domain will produce ringing in the image. Thus, there is a tradeoff between the desire to eliminate aliasing and a desire to have a high MTF. The following metrics consider how uch aliasing can be tolerated for "good" imagery.

11.8.1. LEGAULT CRITERION

According to Legault,[24] when 95% of signal amplitude spectrum is within u_N, aliasing should be negligible:

$$\frac{\int_0^{u_N} O(u) MTF_{SYSTEM}(u)\, du}{\int_0^{\infty} O(u) MTF_{SYSTEM}(u)\, du} > 0.95 \ . \qquad (11\text{-}24)$$

If the scene contains all frequencies, then an optical anti-aliasing filter is needed to meet this criterion. Legault continues *We really need more experimental evidence...95% is a very shaky threshold value... The reader is left to ponder the effects of aliasing further.*

11.8.2. SPURIOUS RESPONSE

Shade[25] introduced the concept of spurious response to quantify the amount of aliasing tolerated when viewing general imagery. Spurious response is applied to vertical direction only because of concerns about raster visibility and its associated aliasing. Recall that early televisions were analog devices in the horizontal direction. The spurious response is simply

$$F_{11}(v) = r_{SP}(v) = MTF_{POST}(v) \sum_{m=1}^{\infty} MTF_{PRE}(mv_S \pm v) O(mv_S \pm v) \ . \qquad (11\text{-}25)$$

Shade only considered the first side band and the worst case where the object consists of all frequencies:

$$r_{SP}(v) = MTF_{POST}(v) \, MTF_{PRE}(v_S - v) \, . \quad (11\text{-}26)$$

According to Shade, aliasing is tolerable when the maximum value of the spurious response is less than 0.15. With most systems, the maximum value of the spurious response occurs at the Nyquist frequency.

Kennedy[26] extended Shade's approach by including the area under the curve. With his approach, the total amount of aliased signal is compared to the direct response:

$$r_{SP\text{-}AREA} = \frac{\int_0^\infty MTF_{POST}(u) \sum_{n=1}^{\infty} MTF_{PRE}(nu_S \pm u) O(nu_S \pm u) \, du}{\int_0^\infty MTF_{SYSTEM} O(u) \, du} \, , \quad (11\text{-}27)$$

where $MTF_{SYSTEM}(u) = MTF_{PRE}(u) MTF_{POST}(u)$. Spurious response is scene dependent. Figure 11-7 illustrates the area-related spurious response for a system that has an ideal reconstruction filter and a practical reconstruction filter.

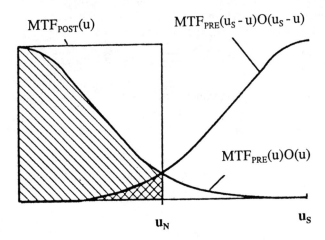

Figure 11-7. The ratio of the shaded areas is the area-related spurious response. (a) Ideal reconstruction filter. (Continued next page).

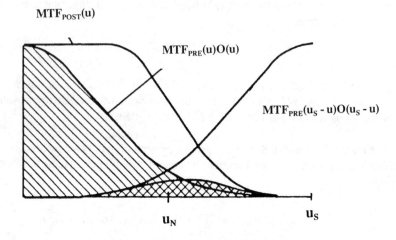

Figure 11-7 (continued). (b) Practical reconstruction filter.

11.8.3. ALIASED SIGNAL as NOISE

A comprehensive end-to-end analysis should include both aliased signal and aliased noise. Park and Hazra[3,4] considered aliased signal to be part of the noise spectrum because it interferes with the ability to perceive targets. Using Equation 11-12 (page 299), the first term is the signal:

$$SIGNAL = MTF_{SYSTEM}(u)\, O(u) \:. \tag{11-28}$$

The second term is signal-dependent (multiplicative) "noise" spectrum:

$$"NOISE" = MTF_{POST}(u) \sum_{n=1}^{\infty} MTF_{PRE}(nu_S \pm u)\, O(nu_S \pm u) \:. \tag{11-29}$$

For a specific frequency, the "image" signal-to-noise ratio is

$$SNR = \frac{MTF_{SYSTEM}(u)\, O(u)}{\int_0^{\infty} MTF_{POST}(u) \sum_{n=1}^{\infty} MTF_{PRE}(nu_S \pm u)\, O(nu_S \pm u)\, du} \:. \tag{11-30}$$

This equation can be modified[3,4] to include detector noise. As with most image quality metrics, quality is scene dependent. This SNR decreases as the aliased signal increases. But a "low" SNR does not mean that the system is unsuitable for a particular application.

11.9. SAMPLING AND RECONSTRUCTION BLUR

As described in Section 8.2., *Edge Ambiguity* (page 201), sampling creates a ghost $\pm\tfrac{1}{2}$ pixel wide in the reconstructed image. The blurriness of an edge depends on the scene sample phase *and* the reconstruction process. Park and Schowengerdt[27] defined the square of the radiometric error as

$$SR = \frac{1}{45\,s^2} \quad \text{when } 1 \leq s \leq 2, \tag{11-31}$$

where s is the number of samples per pixel. At one sample per pixel, the average error is four times higher than at two samples per pixel. They limited their analysis to $1 \leq s \leq 2$ because most scanning systems operate in this region. While Figure 8-8 (page 203) illustrated one sample per pixel, two samples per pixel can be achieved with the scanning geometries shown in Figure 5-19 (page 118). SR is proportional to the shaded area in Figure 8-8.

The image blur decreases when s > 2 but not as fast as implied by Equation 11-29. Park and Schowengerdt's approach is strictly a mathematical method that describes the effects of sampling. It does not consider visual interpretation of image quality.

11.10. NIIRS

The Imagery Interpretability Rating Scale (IIRS) was developed as a numerical scale that measures the potential interpretability of aerial imagery captured on film. It may be considered the information potential for intelligence purposes. It is a set of exploitable tasks (or criteria) equally spaced across a psychophysically defined interpretability scale. Each task consists of three parts: (1) a discrimination level (e.g., detect), (2) an object (e.g., a spare tire), and (3) a modifier (e.g., on a medium-sized truck). As the rating increases, more detail can be extracted from the image. Given an IIRS rating, all lower-valued IIRS tasks can be performed.

312 SAMPLING, ALIASING, and DATA FIDELITY

The 1974 IIRS[28] was significantly updated in 1991 and revised in 1994 to create the National Imagery Interpretability Rating Scale (NIIRS). It is the principle image quality metric used by the reconnaissance community.[29,30] The scale has become an important tool for defining image requirements, selecting and tasking imaging systems, and specifying performance of new systems. The current version replaces outdated equipment references and corrected some errors. Today[31], NIIRSs exist for visible (Table 11-3) and infrared (Table 11-4) sensors reflecting the fact that different sensors highlight different target features. A civilian version also exists (Table 11-5). This new category reflects the transition of aerial imagery from military intelligence gathering into civilian applications. Additional NIIRSs exist for radar and multispectral systems[31]. They are not listed here because that imagery is not typically characterized by sampling effects.

Table 11-3
Visible NIIRS Examples

Rating	Examples
0	Interpretability of imagery precluded by obscuration, degradation, or very poor resolution
1	Detect medium size port facility Distinguish runways at large airfield
2	Detect large hangars at airfields Detect large buildings (hospitals, factories)
3	Detect trains (not individual cars) Identify large surface ship in port
4	Identify large fighter aircraft Identify individual railroad tracks
5	Identify radar as vehicle mounted or trailer mounted Identify individual rail cars (e.g., gondola, flat, box)
6	Identify spare tire on medium-sized truck Identify automobiles as sedans or station wagons
7	Identify ports, ladders, vents on electronic vans Identify individual rail road ties
8	Identify rivet lines on bomber aircraft Identify windshield wipers on a vehicle
9	Identify vehicle registration numbers on trucks Detect individual railroad spikes

When viewing aerial imagery an image analyst determines the smallest discernible cycle on the ground. This cycle width (bar plus space) is the ground resolved distance (GRD). It includes the system MTF, possible degradation by the atmosphere, and system noise. With no degradation, the GRD equals two times the GSD (see Section 10.1.2., *Detector Resolution*, page 277). However, real systems have an MTF and therefore the GRD will be larger than twice the GSD. The relationship between the two is not simple. While the IIRS was based[30] on both the ground resolved distance (Table 11-6) and tasks, the NIIRS is specified by task only.

Table 11-4
Infrared NIIRS Examples

Rating	Examples
0	Interpretability of imagery precluded by obscuration, degradation, or very poor resolution
1	Identify aircraft parking areas at a large airfield Detect industrial area
2	Distinguish between large and small aircraft Detect active stacks on medium-sized surface vessels
3	Detect location of inactive engines on large aircraft Detect individual (inactive) vehicles on a paved area
4	Detect rotor blades on medium-sized helicopter Distinguish individual cars in a train
5	Identify antenna dishes on a radio relay tower Identify missile tubes on submarines
6	Distinguish between small and medium helicopters Identify automobiles as sedans or station wagons
7	Identify exhaust nozzle on an SA-5 missile Identify an individual person
8	Identify antenna on a jeep-sized vehicle Identify equipment in an open-bed, light duty truck
9	Identify door handles on small to medium vehicles Distinguish fingers on a person's hand

Table 11-5
Civilian NIIRS Examples

Rating	Examples
0	Interpretability of imagery precluded by obscuration, degradation, or very poor resolution
1	Distinguish major land classes (urban, forest, etc.) Identify large drainage patterns
2	Identify large fields during the growing season Identify road patterns (clover leafs, etc.)
3	Detect individual houses in a residential area Distinguish natural forest stands from orchards
4	Identify farm buildings as barns, silos, or residences Detect jeep trail through grassland
5	Identify Christmas tree plantations Detect large animals (e.g., elephants, rhinoceros)
6	Detect narcotics intercropping based on texture Detect foot trails through barren areas
7	Identify mature cotton plants in a known cotton field Detect individual steps on a stairway
8	Count individual baby pigs Identify individual pine seedlings
9	Identify individual barbs on a barbed wire fence Identify an ear tag on large game animals

The disappearance of the GRD from the NIIRS was not an oversight. Image interpretability depends on object size, object shape, and contextual cues. For example, with a NIIRS of 8, it should be possible to identify windshield wipers on a vehicle. Here, identification depends on location. The same wipers may not even be detected on the ground. If image quality was related exclusively to the GRD, then the wipers should be identified regardless of location. NIIRS makes sense to the user. It is a measure of system utility. However, system design is based calculable numbers such GSD, MTF, and signal-to-noise ratio.

Table 11-6
IIRS and GRD

IIRS Rating	GRD
0	NA
1	Greater than 9 m
2	4.5 to 9 m
3	2.5 to 4.5 m
4	1.2 to 2.5 m
5	0.75 to 1.2 m
6	40 to 75 cm
7	20 to 40 cm
8	10 to 20 cm
9	Less than 10 cm

The General Image Quality Equation (GIQE) solves this dilemma by bridging the gap between system design parameters and NIIRS. It was developed from a large data base of images evaluated by trained analysts. The GIQE was obtained from a multiple regression analysis. For system operating in the visible, Version 3.0 provides[32]:

$$NIIRS = 11.81 + 3.32\log_{10}\left(\frac{RER_{GM}}{GSD_{GM}}\right) - 1.48 H_{GM} - \frac{G}{SNR} \quad (11\text{-}32)$$

where
  GSD is the ground sampled distance measured in inches
  RER is the slope of the relative edge response
  H is the overshoot due to edge sharpening (MTF compensation)
  G is the noise gain due to edge sharpening
  SNR is the signal-to-noise ratio
Because $1/\log(2) = 3.32$, some authors express the GIQE with $\log(2)$.

316 SAMPLING, ALIASING, and DATA FIDELITY

The subscript GM indicated that the values are the geometric mean of the vertical and horizontal components. The geometric GSD is

$$GSD_{GM} = \sqrt{GSD_H \, GSD_V} \quad \text{inches} \;, \quad (11\text{-}33)$$

where

$$GSD_H = \frac{d_{CCH}}{fl}(RANGE) \quad \text{and} \quad GSD_V = \frac{d_{CCV}}{fl}(RANGE) \;. \quad (11\text{-}34)$$

In the horizontal direction, the relative edge response is

$$ER_H(a) = \frac{1}{2} + \frac{1}{\pi} \int_0^\infty \frac{MTF_{ATM}(u) \, MTF_{SYSTEM}(u)}{u} \sin(2\pi a u) \, du \;, \quad (11\text{-}35)$$

where u is the spatial frequency in cycles per sample spacing or, equivalently, $u_i d_{CCH}$. For the vertical direction, u is replaced with v. The value a is the measured from the edge transition. The edge response is normalized so that as a becomes large, ER(a) approaches unity. The slope is calculated from the difference in the response at a = 0.5 and a = -0.5 (Figure 11-8). Although the GIQE includes atmospheric turbulence and boundary wavefront error MTFs, the atmospheric effects are often omitted ($MTF_{ATM}(u) = 1$). The geometric RER is

$$RER_{GM} = \sqrt{[ER_H(0.5) - ER_H(-0.5)][ER_V(0.5) - ER_V(-0.5)]} \;. \quad (11\text{-}36)$$

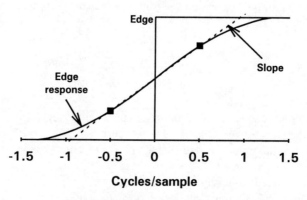

Figure 11-8. Relative edge response as a function of the cycles per sample (u = $u_i d_{CC}$). For the horizontal direction, the edge is vertically oriented. The squares indicate ER(-0.5) and ER(0.5). A similar response occurs in the vertical direction where the edge is horizontally oriented.

H is the edge response at a = 1.25. When MTF boost is present, the edge response will typically peak 1 to 3 units from the transition (Figure 11-9). Then H is the maximum value of this response. The geometric mean is

$$H_{GM} = \sqrt{H_H H_V} \ . \qquad (11\text{-}37)$$

Because boost degrades the NIIRS value (provides a smaller value), most reconnaissance systems are built without boost. For a detector-limited system (optics cutoff is 5× detector cutoff), RER ≈ 0.93 and H ≈ 1.0. When the display is included, RER ≈ 0.57 and H ≈ 1.0. This assumes a 100% fill-factor array.

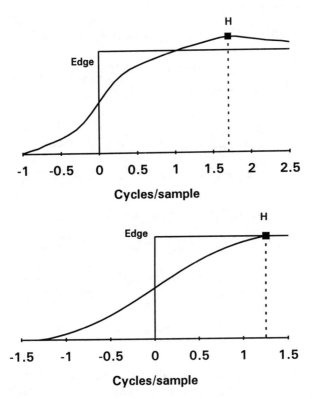

Figure 11-9. MTF compensation as a function of the cycles per sample ($u = u_i d_{CC}$). When peaking occurs, H is the maximum value of the edge between a = 1 and a = 3. With a monotonically increasing edge function, H is evaluated at a = 1.25.

318 SAMPLING, ALIASING, and DATA FIDELITY

The noise gain is

$$G = \sqrt{\sum_{1}^{m}\sum_{1}^{n}(KERNEL_{ij})^2} , \qquad (11\text{-}38)$$

where the gain coefficients of the boost are in the kernel encompassing m × n pixels (assuming one-to-one mappping from pixels to datels). With no boost, G = 1.

While Version 3.0 predicted NIIRS when RER > 0.9 and H ≈ 1.0, Version 4.0[33] covers a wider range:

$$NIIRS = 10.251 - k_1 \log_{10}(GSD_{GM}) + k_2 \log(RER_{GM})$$
$$- 0.656 H_{GM} - 0.334 \frac{G}{SNR} . \qquad (11\text{-}39)$$

The coefficients are

$$k_1 = 3.32 \text{ and } k_2 = 1.559 \text{ when } RER \geq 0.9$$
$$k_1 = 3.16 \text{ and } k_2 = 2.817 \text{ when } RER < 0.9 . \qquad (11\text{-}40)$$

For system operating in the infrared, Version 4.0 provides:

$$NIIRS = 9.751 - k_1 \log_{10}(GSD_{GM}) + k_2 \log(RER_{GM})$$
$$- 0.656 H_{GM} - 0.334 \frac{G}{SNR} . \qquad (11\text{-}41)$$

The only difference between the visible and infrared equations is the first constant.

Although NIIRS does not include the GRD, there is a tendency for the analyst to calculate the NIIRS value and use Table 11-6 to estimate the GRD. This estimation is not discussed in the GIQE documentation and, therefore, not recommended.

IMAGE QUALITY METRICS

Finally, NIIRS is used as a design tool for high altitude aerial imagery. Sensors mounted on ground vehicles or low flying aircraft use probability of discrimination as a design guide (see Section 11.7., *MRT and MRC*, page 305). The probability of detection, recognition, and identification can be related to the number of cycles, N_{CYCLE}, across a target's critical dimension, $D_{CRITICAL}$. The critical dimension is the geometrical mean of the height and width of a rectangular target. Then

$$N_{CYCLE} = \frac{D_{CRITCAL}}{2 \, GSD} . \qquad (11\text{-}42)$$

Solving Equation 11-32 or Equation 11-39 for the GSD and substituting into Equation 11-42 provide a relationship between N_{CYCLE} and NIIRS. By using Equation 11-22 (page 307), the performance probability can be found as a function of the NIIRS value[34-36].

11.11. GENERAL TRENDS

The image quality metrics presented assumed that the system was linear. Some modifications were employed to incorporate sampling effects. However, most systems today use nonlinear digital filters. Data analysis techniques now include the neural net approach and fuzzy logic. High definition television and image storage techniques use image compression algorithms to conserve bandwidth or storage space. The effects of these techniques cannot be described in closed form. Rather, representative images are evaluated to determined the effectiveness of the filters, compression or image processing techniques.[37]

Pitas and Venetsanopoulos[38] describe the effectiveness of 29 filters (median, harmonic mean, dilation, etc.) on different types of noise (Gaussian, salt and pepper, etc) and different image characteristics (edge preservation, detail preservation, etc.). However, they provided only general guidelines by specifying effectiveness as poor, average, or good. They admit that this is a crude ranking. Specific images can be selected when a particular filter performs extremely well and others where performance is unacceptable.

Weeks[39] provides limited imagery illustrating compression and the error introduced with compression

$$f_{ERROR}(x,y) = f_{UNCOMPRESSED}(x,y) - f(x,y) . \qquad (11\text{-}43)$$

It is implied that as the $f_{ERROR}(x,y)$ increases in intensity, the compression-decompression algorithm is less desirable. This should not be interpreted as meaning that the decompressed image cannot be used for a particular application.

The number of nonlinear filter algorithms are increasing faster than our ability to upgrade image quality metrics. Given resource constraints (time, money, and personnel), nonlinear filter effectiveness will probably be demonstrated on a minimal number of images. These include the popular three- or four-bar test patterns, Lena, and the mandrill. The extension to any other image is pure conjecture.

11.12. REFERENCES

1. L. C. Biberman, *Editor's Introduction*, in *Perception of Displayed Information*, L. C. Biberman, ed., pp. 233-237, Plenum Press, New York (1973).
2. R. Legault, "The Aliasing Problems in Two-dimensional Sampled Imagery," in *Perception of Displayed Information*, L. C. Biberman, ed., pp. 292-295, Plenum Press, New York (1973).
3. S. K. Park and R. Hazra, "Aliasing as Noise: A Quantitative and Qualitative Assessment," in *Infrared Imaging Systems: Design, Analysis, Modeling and Testing IV*, G. C. Holst, ed., SPIE Proceedings Vol. 1969, pp. 54-65 (1993).
4. S. K. Park and R. Hazra, "Image Restoration Versus Aliased Noise Enhancement," in *Visual Information Processing III*, F. O. Huck and R. D. Juday, eds., SPIE Proceedings Vol. 2239, pp. 52-62 (1994).
5. A. R. Weeks, Jr., *Fundamentals of Electronic Image Processing*, pp. 228-293, SPIE Optical Engineering Press, Bellingham, WA (1996).
6. A. R. Weeks, Jr., *Fundamentals of Electronic Image Processing*, pp. 90-120, SPIE Optical Engineering Press, Bellingham, WA (1996).
7. S. K. Park, "Image Gathering, Interpolation, and Restoration: A Fidelity Analysis," in *Visual Information Processing*, F. O. Huck and R. D. Juday, eds., SPIE Proceedings Vol. 1705, pp. 134-144 (1992).
8. H. Kusaka, "Consideration of Vision and Picture Quality - Psychological Effects Induced by Picture Sharpness," in *Human Vision, Visual Processing, and Digital Display*, B. E. Rogowitz, ed., SPIE Proceedings Vol. 1077, pp. 50-55 (1989).
9. O. H. Shade, Sr., "Image Gradation, Graininess, and Sharpness in Television and Motion Picture Systems," published in four parts in *J. SMPTE*: "Part I: Image Structure and Transfer Characteristics," Vol. 56(2), pp. 137-171 (1951); "Part II: The Grain Structure of Motion Pictures - An Analysis of Deviations and Fluctuations of the Sample Number," Vol. 58(2), pp. 181-222 (1952); "Part III: The Grain Structure of Television Images," Vol. 61(2), pp. 97-164 (1953); "Part IV: Image Analysis in Photographic and Television Systems," Vol. 64(11), pp. 593-617 (1955).
10. H. L. Snyder, "Image Quality and Observer Performance," in *Perception of Displayed Information*, L. M. Biberman, ed., pp. 87-118, Plenum Press, New York, NY (1973).
11. F. J. Drago, E. M. Granger, and R. C. Hicks, "Procedures for Making Color Fiche Transparencies of Maps, Charts, and Documents," *Journal of Imaging Science and Technology*, Vol. 11(1) pp. 12-17 (1965).

12. E. M. Granger and K. N. Cupery, "An Optical Merit Function (SQF) Which Correlates With Subjective Image Judgments," *Photographic Science and Engineering*, Vol. 16, pp. 221-230 (1972).
13. N. Nill, "A Visual Model Weighted Cosine Transform for Image Compression and Quality Measurements," *IEEE Transactions on Communications*, Vol. 33(6), pp. 551-557 (1985).
14. T. J. Schultz, "A Procedure for Calculating the Resolution of Electro-Optical Systems," in *Airborne Reconnaissance XIV*, P A. Henkel, F. R. LaGesse, and W. W. Schurter, eds., SPIE Proceedings Vol. 1342, pp. 317-327, (1990).
15. A. N. deJong and S. J. M. Bakker, "Fast and Objective MRTD Measurements," in *Infrared Systems - Design and Testing*, P. R. Hall and J. S. Seeley, eds., SPIE Proceedings Vol. 916, pp. 127-143 (1988).
16. F. W. Campbell and J. G. Robson, "Application of Fourier Analysis to the Visibility of Gratings," *Journal of Physiology*, Vol. 197, pp. 551-566 (1968).
17. P. G. Barten, "Evaluation of Subjective Image Quality with the Square-root Integral Method," *Journal of the Optical Society of America. A*, Vol. 17(10), pp. 2024-2031 (1990).
18. P. G. Barten, "Evaluation of the Effect of Noise on Subjective Image Quality," in *Human Vision, Visual Processing and Digital Display II*, J. P. Allenbach, M. H. Brill, and B. E. Rogowitz, eds., SPIE Proceedings Vol. 1453, pp. 2-15 (1991).
19. P. G. Barten, "Physical Model for the Contrast Sensitivity of the Human Eye," in *Human Vision, Visual Processing, and Digital Display III*, B. E. Rogowitz, SPIE Proceedings Vol. 1666, pp. 57-72 (1992).
20. P. G. Barten presents short courses at numerous symposia. See, for example, "Display Image Quality Evaluation," *Application Seminar Notes*, SID International Symposium held in Orlando, Fl (May 1995). Published by the Society for Information Display, Santa Ana, CA, or "MTF, CSF, and SQRI for Image Quality," IS&T/SPIE's Symposium on Electronic Imaging: Science & Technology, San Jose CA (February 1995).
21. G. C. Holst, *Electro-Optical Imaging System Performance*, pp. 381-411, JCD Publishing, Winter Park, FL (1996).
22. G. C. Holst, *CCD Arrays, Cameras, and Displays*, pp. 293-324, JCD Publishing, Winter Park, FL (1996).
23. G. C. Holst, *Electro-Optical Imaging System Performance*, pp. 456-457, JCD Publishing, Winter Park, FL (1996).
24. R. Legault, "The Aliasing Problems in Two-dimensional Sampled Imagery," in *Perception of Displayed Information*, L. C. Biberman, ed., page 305, Plenum Press, New York (1973).
25. O. H. Shade, Sr., "Image Reproduction by a Line Raster Process," in *Perception of Displayed Information*, L. C. Biberman, ed., pp. 233-278, Plenum Press, New York (1973).
26. H. V. Kennedy, "Modeling Second-generation Thermal Imaging Systems," *Optical Engineering* Vol. 30(11), pp. 1771-1778 (1991).
27. S. K. Park and R. A. Schowengerdt, "Image Sampling, Reconstruction and the Effect of Sample-scene Phasing," *Applied Optics*, 21(17), pp. 3142-3151 (1982).
28. K. Riehl and L. Maver, "A Comparison of Two Common Aerial Reconnaissance Image Quality Measures," in *Airborne Reconnaissance XX*, W. G. Fishell, A. A. Andraitis, A. C. Crane, Jr., and M. S. Fagan, eds., SPIE Proceedings Vol. 2829, pp. 242-254 (1996).
29. K. Riehl, "A Historical Review of Reconnaissance Image Evaluation Techniques," in *Airborne Reconnaissance XX*, W. G. Fishell, A. A. Andraitis, A. C. Crane, Jr., and M. S. Fagan, eds., SPIE Proceedings Vol. 2829, pp. 322-334 (1996).
30. Air Standardization Agreement: "Imagery Interpretability Rating Scale," Air Standardization Co-ordinating Committee report AIR STD 101/11 dated 10 July 1978.
31. J. M. Irvine, "National Imagery Interpretability Rating Scales (NIIRS): Overview and Methodology," in *Airborne Reconnaissance XXI*, W. C. Fishell, ed., SPIE Proceedings Vol. 3128 (1997).

32. "General Image Quality Equation - User's Guide, Version 3.0," HAE UAV Tier II+ (1994).
33. "General Image Quality Equation - User's Guide, Version 4.0," National Imagery and Mapping Agency, Data and Systems Division, Springfield, VA (1996).
34. R. Driggers, M. Kelley, and P. Cox, "National Imagery Interpretability Rating Scale and the Probabilities of Detection, Recognition, and Identification," *Optical Engineering*, Vol. 36(7), pp. 1952-1959 (1997).
35. R. Driggers, P. Cox, J Leachtenauer, R. Vollmerhausen, and D. A. Scribner, "Targeting and Intelligence Electro-Optical Recognition Modeling: A Juxtaposition of the Probabilities of Discrimination and the General Image Quality Equation,", to appear in *Optical Engineering* (1998).
36. R. E. Hanna "Using GRD to Set EO Sensor Design Budgets," in *Airborne Reconnaissance XXI*, W. C. Fishell, ed., SPIE Proceedings Vol. 3128 (1997).
37. See, for example, *Very High Resolution and Quality Imaging II*, V. R. Algazi, S. Ono, and A. G. Tescher, eds., SPIE Proceeding Vol. 3025 (1997).
38. I. Pitas and A. N. Venetsanopoulos, *Nonlinear Digital Filters*, pp. 325-328, Kluwer Academic Publishers, Boston, MA (1990).
39. A. R. Weeks, Jr., *Fundamentals of Electronic Image Processing*, pp. 471-547, SPIE Optical Engineering Press, Bellingham, WA (1996).

INDEX

Aerial image modulation 284, 302
Airy disk 27, 70, 253, 277
Aliased noise 310
Aliasing 14, 89, 166, 193, 215, 261, 294
Ambiguity 11, 202
Amplitude spectrum 33
Analog multiplexer 99
Analog-to-digital converter 102
Anti-alias filter 14, 91, 125, 240
    optical prefilter 123
Aperture correction 243
Averaging filter 138

Band-limited 81
Bar pattern 62, 215
Beat frequency 211
Binary 5
Blockiness 109, 165
Blur 205, 297
Blur diameter 277
Bomber 221
Boost filter 243
Butterworth filter 241

Causality 30, 65, 132
CCD 110, 203, 231
CCD camera 226, 250, 294
Center-to-center spacing 26, 96, 107
Character recognition 224
Chebyshev filter 162, 242
Collimator 231
Color CRT 179
Color filter array 110
Compact disk 167
Compression 319
Continuous/continuous model 20
Continuous/discrete model 22
Continuous/discrete/continuous model 22
Contrast sensitivity 251
Contrast transfer function 220
Convolution 59, 67, 70
Critical dimension 306
CRT 160
    color 179
    digital 179
    raster scanned 173
    resolution 281
    spot size 173
Cubic spline 151
Cutoff
    detector 105, 237
    optical 79, 235

DAS 278
Data acquisition 15
Datel 24
Decimation 136
Demand modulation function 302
Detection 302, 306, 319
Detector 103
    cutoff 105, 237
    MTF 237
    pitch 107
Detector-angular-subtense 26, 278
Detector-limited 203, 254, 278
DFT 30
Digital display 179
Digital filter 245
Digital number 5
Digital-to-analog converter 82, 161
Dirac delta 43, 67
Discrete Fourier transform 30, 44
Discrete/discrete model 20
Disel 24
Display 171, 281, 297
Display MTF 249
Display space 230
Display-limited 160
Distortionless transmission 77
Dither 113
Dixel 24
Dots per inch 224
Dwell 121
Dynamic sampling 226
Dynamic scene projector 269

Edge location 201
Edge sharpness 166
Effective-instantaneous-field-of-view 278
EIA RS-170 132, 155, 233, 281, 289
Electronic resolution 281
Electronic transfer function 31
Equivalent pass band 286, 301

F-number 279
Fax machine 132, 160
FFT 30
Fill-factor 26, 237, 278
Filter
    anti-alias 14, 91, 125, 240
    averaging 138, 247
    boost 243
    digital 245
    first-order 164
    ideal 77, 87, 162
    low-pass 12, 162, 240
    optical prefilter 123
    reconstruction 12, 87, 240, 264
    sample-and-hold 12
    spatial 30
    time 30
    zero-order 12, 83, 88, 163
Finite impulse response 245
Flash converter 102
Flat bed scanner 132
Flat field 175
Flat panel display 160, 181, 251
Flaw 205
Font size 224
Four-bar pattern 64, 220, 227
Fourier integral 39
Fourier series 31
Fourier transform 40
    discrete 44
    fast 30
Frame grabber 102, 129, 154, 289
Frequency scaling 264
Frequency shift 142
Frequency spectrum 33
Fuzzy logic 319

General Image Quality Equation 315
Gibbs phenomenon 38, 80, 152, 265
GIQE 315
GRD 279, 313
Ground resolved distance 279, 313
Ground sampled distance 280, 315
GSD 280, 315

Halftone imagery 185
Halftoning 160
Hamming window 61
Hanning window 61
Hardware-in-the-loop 19, 269
Hexadecimal 5

Hexagonal sampling 126
Horizontal resolution 282
Human visual system 168
Human visual system MTF 251

Ideal filter 77, 87, 162
Idealized circuits 77
Idealized components 77
Identification 221, 306, 319
IIRS 279, 311
Image enhancement 16
Image fidelity 295
Image quality metrics 297
Image restoration 17
Image space 230
Image warping 131
Imagery Interpretability Rating
    Scale 279, 311
Impulse 67
Impulse function 43
In-phase 197, 238
Infinite impulse response 245
Infrared image 269
Infrared system 227
Instantaneous-field-of-view 278
Intermediate spectra 139
Interpolation 139, 144

Kell factor 281

Landsat 225
Laplace transform 78
Laser printer 131, 183
Leakage 59
Legault criterion 308
Limiting resolution 283
Line spread function 70
Linear system theory 65
Linear-shift-invariant 65
Linearity 42
Low-pass filter 12, 162, 240

Machine vision 19, 160, 224, 225, 269
Microdither 113
Microlens 125
Microscan 113, 226
Military system 300
Minimum dimension 205
Minimum resolvable contrast 306
Minimum resolvable temperature 306
Missile seeker 269

INDEX 325

Model, system 20
Modulation 73
  threshold 168, 251
Modulation transfer function 71, 73
Modulation transfer function area 302
Moiré 18, 96
Monitor 233
MTF 229
  averaging filter 247
  boost 243
  Butterworth 241
  Chebyshev 242
  CRT 249
  detector 237
  diffraction-limited 79
  digital filter 247
  display 249
  flat panel 251
  human visual response 251
  lens 79
  optical 235
  sample-scene 238
  system 253
  tuned circuit 243
  unsharp mask 245
Multiplexer 99
Multiplexer, analog 99

National Imagery Interpretability
  Rating Scale 279, 312
Neural net 319
NIIRS 279, 312
NTSC 155, 233, 281
Nyquist frequency 2, 110, 279

Object space 230
Observer space 230
Observer-limited 160
Octal 5
Optical character reader 224
Optical MTF 235
Optical prefiltering 123, 227
Optical resolution 277
Optical transfer function 31, 71
Optics-limited 203, 254, 278
Out-of-phase 197, 238

PAL 155, 233
PAS 26, 121, 278

Pattern
  bar 215
  four-bar 62, 217, 227
  sweep frequency 169, 220
  three-bar 62
Pel 24
Phase shift 77
Phase spectrum 33
Phase transfer function 31, 71
Phasing 223
Phasing effect 197
Photographic camera 231
Pitch 26, 107, 279
Pixel 24
Pixel-angular-subtense 26, 121, 278
Pixellation 88, 186
Pixels on target 221
Point spread function 70
Power 33, 41
Prefiltering 123

Quantization 5

Raised cosine window 61
Raster scanned device 173
Rayleigh criterion 277, 297
Recognition 221, 306, 319
Reconnaissance 312
Reconstruction 12, 87, 158
Reconstruction filter 87, 162, 240, 264
Remapping 133
Remote sensing 133
Resampling 131
Resel 24, 277
Resolution 275, 297
  CRT 281
  detector 278
  electronic 281
  horizontal 282
  limiting 283
  optical 277
  Séquin 285
  Shade 286
  TV limiting 282
  vertical 281, 285
Resolution/addressability ratio 172, 224, 283
Ringing 38, 80, 91, 152, 166, 243, 265

RS-170   121, 132, 155, 233, 281, 289

Sample-scene MTF   238
Sampled data systems   82
Sampling and reconstruction blur   311
Sampling devices   98
Sampling theorem   81
Scaling   42
Scanning system   119
Scene generator   269
Scene projector   269
Scenel   24
SECAM   233
Sensor fusion   133
Séquin's resolution   285
Shade's equivalent pass band   286, 301
Shade's equivalent resolution   286
Shannon's sampling theorem   81
Signal-to-noise ratio   305
Simulation   260
Sinc function   34
Sine wave response   251
Spatial filter   30
Spatial frequency   230
Spline function   147
Spurious frequency   165
Spurious response   308
Square wave   35, 79, 170, 196, 201, 215
Square-root integral   305
Staring array   115
Strehl ratio   297
Subjective quality factor   302
Superposition   67, 78
Sweep frequency pattern   169, 220
System model   20

Taps   247
Target
    four-bar   62, 220, 227
    sweep frequency   169, 220
    three-bar   62, 220, 227
Television   173, 294
Three-bar pattern   62, 220, 227
Threshold   206
Threshold modulation   168, 251, 283, 302
Time filter   30
Translation   42
TV limiting resolution   282
Twinkle   109

Undersample   90, 96
Unrealizable components   77
Unsharp mask   189, 245

Vertical resolution   281
Video
    bandwidth   154
    standard   154
Video standard   233
Video timing   233

Window   59
    Hamming   61
    Hanning   61
    raised cosine   61

Zoom
    electronic   186
    optical   189